Implementation and Evaluation of
Unmanned Aerial Vehicles and
Sensor Systems in Weed Research

Implementation and Evaluation of Unmanned Aerial Vehicles and Sensor Systems in Weed Research

Dissertation to obtain the doctoral degree of Agricultural Sciences (Dr. sc. agr.)

Faculty of Agricultural Sciences
University of Hohenheim

Intitute of Phytomedicine
Department of Weed Science

submitted by

Robin Mink
from *Stuttgart*
in *2019*

Bibliografische Information der Deutschen Nationalbibliothek

Die Deutsche Nationalbibliothek verzeichnet diese Publikation in der
Deutschen Nationalbibliografie; detaillierte bibliographische Daten sind im Internet
über http://dnb.d-nb.de abrufbar.
1. Aufl. - Göttingen: Cuvillier, 2020
Zugl.: Hohenheim, Univ., Diss., 2019

D100

Abbildung Cover: © Jannis Machleb

ISBN 978-3-7369-7178-3
eISBN 978-3-7369-6178-4

Zusammenfassung

Eine wirkungsvolle Unkrautbekämpfung ist seit jeher ein zentraler Bestandteil des Kulturpflanzenanbaus. Seit der Erfindung und Kommerzialisierung des chemischen Pflanzenschutzes nimmt jedoch die Kritik gegenüber der vermehrten Anwendung von Herbiziden stetig zu. Die ackerbaulichen Nachteile sowie die Externalitäten des umfangreichen Herbizideinsatzes liegen in den damit verbundenen Umweltbelastungen und der Selektion herbizidresistenter Unkräuter. Mit dem Aufkommen von Precision Farming Technologien wurden vermehrt Sensoren und Aktoren in diversen landwirtschaftlichen Maschinen implementiert. Im Pflanzenschutz unterstützen Precision Farming Technologien, Landwirte und Wissenschaftler bei der präzisen Durchführung und Überwachung von Unkrautbekämpfungsmaßnahmen. Die Weiterentwicklung der sensorgestützten Informationstechnologie in der Herbologie bot neue Möglichkeiten Unkrautmanagementstrategien hinsichtlich ihrer Wirksamkeit und Nachhaltigkeit zu verbessern. Der Fokus dieser Dissertation lag auf der Entwicklung und Bewertung von Methoden zur Reduktion des Herbizideinsatzes, sowie auf der sensorgestützten Überwachung der Reaktion von Kulturpflanzen und Unkräutern nach einer Herbizidapplikation. Besonderes Augenmerk wurde auf die Entwicklung von Erfassungsmethoden mit unbemannten Luftfahrzeugen (UAVs), als Sensorträger, und bodengestützten Sensoren gelegt, um die folgenden Forschungsziele zu erarbeiten:

- Entwicklung und Validierung von UAV-basierten Applikationskarten für teilschlagspezifische Herbizidanwendungen.

- Bewertung des Herbizidstresses in Mais durch bodennahe Sensormessungen und Luftbilder.

- Training und Validierung von Klassifikatoren zur sensorgestützten Erkennung von herbizidresistenten Unkräutern.

In der **1. Studie** wurden Feldversuche zur Erfassung der Unkrautverteilung mittels hyperspektralen und rot-grün-blau (RGB) Luftbildern durchgeführt. Die aus den Luftbildern berechneten Unkrautverteilungskarten für *Cirsium arvense* und *Rumex crispus* wurden anschließend verwendet, um eine teilschlagspezifische

Herbizidapplikation durchzuführen. Die Berechnung der Herbizidapplikationskarten erfolgte aus den Luftbildern in Kombination mit weiteren Daten aus anderen Precision Farming Anwendungen. Der entwickelte Geobildauswertungsalgorithmus zeigte Identifikationsgenauigkeiten der *Cirsium arvense* und *Rumex crispus* Nester von 96 % in Mais und 90 % in Zuckerrüben. Durch die Anwendung dieser Karten waren Herbizideinsparungen von 43 % bis 96 % in Abhängigkeit von der Unkrautdichte möglich.

In der **2. Studie** wurde eine mögliche Reduktion der Aufwandmenge des Safeners *Cyprosulfamid* in Herbizidanwendungen mit dem Wirkstoff *Isoxaflutol* in Mais untersucht. Die Erfassung der Kulturpflanzenverträglichkeit erfolgte dabei mittels bodennahem Hyperspektralsensor und multispektralen Luftbildaufnahmen, welche mit klassischen Messungen der Bodenwahrheit (Biomasseschnitt und Bonitur der Pflanzenschädigung) verglichen wurden. Insbesondere Maisherbizide zur Nachauflaufbehandlung weisen eine ausreichende Unkraut-/Kulturpflanzenselektivität nur in einem vergleichbar kurzen Vegetationsstadium und oftmals nur in Kombination mit einer Safenerbehandlung auf. In Gewächshaus- und Feldexperimenten wurde die Applikation von *Cyprosulfamid* als Saatgutbeizung vor dem Herbizideinsatz und als Kombinationspräparat mit dem Herbizid in drei aufeinander folgenden Entwicklungsstadien untersucht. Die erfassten Multispektralbilder und Hyperspektralmessungen der Umgebungslichtreflexion der Kulturpflanzen zeigten vergleichbare Ergebnisse mit einer stark positiven Korrelation, sowie eine Vergleichbarkeit mit der Bodenwahrheit. Die sensorbasierten Auswertungen überstiegen dabei die von einem Boniteur beeinflussten visuellen Schätzungen, sowohl in der Erhebungsgeschwindigkeit, als auch im Signifikanzniveau, aufgrund der höheren Konformität der Sensormessungen. Die Gesamtmenge des für eine ausreichende Kulturpflanzenverträglichkeit einer *Isoxaflutol*-Behandlung in Mais benötigten Safeners (*Cyprosulfamid*), konnte durch dessen Anwendung als Saatgutbeizung um 88 % reduziert werden.

Die Entwicklung schneller und zuverlässiger Testsysteme zur Identifizierung herbizidresistenter Unkräuter ist nach wie vor von großer Bedeutung, um eine effiziente Bekämpfung dieser Biotypen noch vor der Samenabreife zu ermöglichen. In der **3. und 4. Studie** wurden zwei sensorgestützte Klassifikatoren zur Unterscheidung untersuchter Biotypen in die Klassen „resistent" und „sensitiv" für Acetolactat-Synthase (ALS)-Inhibitoren resistente *Papaver rhoeas* und *Stellaria media* Pflanzen entwickelt und getestet. In der 3. Studie wurde ein tragbarer Chlorophyllfluorometer verwendet, um einen Maximum-Likelihood-Klassifikator mit Messungen ALS-inhibitor sensitiver und resistenter *Papaver rhoeas* und *Stellaria media* Pflanzen zu trainieren und zu validieren. In der 4. Studie wurde ein künstliches neuronales Netzwerk mit Hyperspektralmessungen der Umgebungslichtreflexion von *Stellaria media* Pflanzen trainiert und validiert. Der Maximum-Likelihood-Klassifikator zeigte bereits innerhalb von 3 Tagen nach der Herbizidbehandlung Klassifizierungsgenauigkeiten von 62 % bis 100 %. Die Klassifizierungsgenauigkeit des künstlichen neuronalen Netzes, zwischen resistenten und sensitiven Pflanzen, lag vier Tage nach der Herbizidbehandlung zwischen 91 % und 95 %. Zusätzlich konnten die Gewichte des ersten Hidden-Layers des künstlichen neuronalen Netzwerks Aufschluss über die Klassifikationsrelevanz der 65 untersuchten Lichtreflexionsspektren geben. Es konnte nachgewiesen werden, dass die identifizierten wichtigen Spektren wesentlich zur Erkennung resistenter Biotypen beitragen.

Die in dieser Dissertation entwickelten Methoden zeigen das Potenzial von UAVs und Sensoranwendungen für die herbologische Untersuchungen zur Erfassung des Kulturpflanzenstresses nach Herbizidapplikationen, die Berechnung von Unkrautverteilungskarten und den damit möglichen Herbizideinsparungen, sowie den Nachweis von Herbizidresistenzen in den untersuchten Unkrautarten.

Summary

Effective weed control in field crops has always been a major challenge in plant cultivation. Since the invention of chemical crop protection, several concerns on the excessive use of herbicides have arisen. The main drawbacks of chemical weed control are the associated environmental impacts and the selection of herbicide-resistant weeds. With the advent of precision farming technologies, sensors and actuators were implemented in agricultural machinery to assist farmers and scientists to monitor and execute weed control measures accurately. The evolution of sensor-based information technology in weed science offered new possibilities to improve weed management strategies. This work aimed at the implementation and evaluation of sensor systems in weed science in order to investigate the possible reduction of weed control related agrochemicals as well as measuring the responses of crops and weeds to herbicide applications. Special attention was paid to the evaluation of detection methods using ground-based sensors and unmanned aerial vehicles (UAVs) as carriers for hyperspectral and red-green-blue (RGB) digital cameras, to investigate the following objectives:

- Development and validation of UAV imagery-based application maps for site-specific herbicide applications.

- Evaluation and comparison of aerial and ground-based sensor measurements to auditor estimates for the assessment of maize stress after herbicide applications.

- Training and validation of classifiers using sensor derived plant information to classify herbicide resistant weed biotypes and identify relevant data inputs.

In the **1st study** field experiments were conducted to collect hyperspectral and red-green-blue (RGB) aerial imagery in order to precisely asses the weed distribution for the calculation of weed maps. The maps were subsequently used to perform a site-specific herbicide application using a real-time-kinematic global navigation satellite system controlled sprayer. The herbicide application maps for *Cirsium arvense* and *Rumex crispus* were created for sugar beet and maize using aerial imagery in combination with comprehensive precision farming information of the field. The developed geo-image analysis algorithm showed identification accuracies of the *Cirsium arvense* and *Rumex crispus* patches of 96 % in maize and 90 % in sugar beet. The UAV based prescription maps enabled herbicide savings of 43 % to 96 %.

The **2nd study** aimed at the reduction of safeners as herbicide partners and a sensor-based evaluation of the maize and weed responses to the examined safener treatments. Especially post-emergent maize herbicides show a sufficient crop/weed selectivity only in a specific time slot and in combination with a safener treatment. In greenhouse and field experiments, the safener was sprayed as a commercial herbicide premix formulation and as seed treatment prior to the herbicide application at three consecutive crop development stages. The maize response after herbicide application was investigated by multispectral airborne imagery and ground-based hyperspectral reflectance measurements. The multispectral imagery and hyperspectral sensor measurements showed similar results with a strong positive correlation and comparability to the ground-truth measurements. The sensor-based evaluations exceeded the auditor affected visual estimations in time and significance levels due to the higher conformity of the results. The total amount of safener (*cyprosulfamide*) needed, for safening a *isoxaflutole* treatment in maize could be reduced by 88 % when applied as a seed treatment, without significant yield losses.

The development of quick and reliable testing systems to detect herbicide resistant weeds is still of great importance, to enable efficient in-season weed control measures and to avoid further field infestations with herbicide resistant weeds. Two sensor-based classifiers were developed and tested for the separation into the classes herbicide 'resistant' and 'sensitive' in two dicotyledonous weed species with *acetolactate*-synthase inhibitors in the **3rd and 4th study**. A chlorophyll fluorometer was used to train a discriminant maximum likelihood classifier with *Papaver rhoeas* and *Stellaria media*, and an artificial neural network was trained on *Stellaria media* hyperspectral plant reflection. The discriminant maximum likelihood classifier showed classification accuracies of 62 % to 100 % already within three days after the herbicide treatment. The separation of resistant and sensitive plants in the artificial neural network showed accuracies of 91 % to 95 % four days after the herbicide treatment. The weights of the first hidden neuron layer of the artificial neural network for the classification of the hyperspectral data were analysed and provided information on the importance of the input spectra, which can aid the selection of suitable sensors. It was proven that the identified important spectra contribute significantly to the recognition of resistant biotypes.

The findings of these studies highlight the potential of the investigated UAVs and sensor applications for weed research-related field surveys by enabling stress estimations, herbicide savings and herbicide resistance detection.

Contents

Chapter 1

General Introduction

Weed control in crop production systems is not only a major challenge in plant cultivation but also an ever-expanding discipline in farming practice and science. An effective weed control in arable farming is needed to ensure yield stability and product quality (Cousens, 1985). Chemical weed control provides enormous cost-effectiveness but at the same time it has effectuated new problems, which may jeopardise the long-term operating efficiency (Kudsk and Streibig, 2003). The major drawbacks are on the associated environmental impacts and the selection of herbicide-resistant weeds (Heap, 2019; Kudsk and Streibig, 2003). In order to respond to the growing awareness of these problems, chemical weed control strategies are increasingly being adapted to the variable conditions of the fields (Gerhards and Oebel, 2006; López-Lozano *et al.*, 2010; Lindblom *et al.*, 2017; Fernandez-Quintanilla *et al.*, 2018; Lambert *et al.*, 2018). Nevertheless, effective weed control remains the primary objective to maintain stable crop yields (Cousens, 1985).

To perpetuate high production standards and concurrently enabling sustainable cropping systems, weed research focuses on the development of sound ecological and environmentally acceptable weed control practises. In order to adopt traditional weed control measures and develop new weed control strategies intensive observations of the crop and weed responses and accompanying effects are required. For the evaluation of the investigated measures, sensor systems are suitable tools to collect precise and quantifiable data (Peteinatos *et al.*, 2014, 2016). Unmanned aerial vehicle (UAV) mounted sensors open up new perspectives on plant populations, which are not comparable with ground-based data collection (Torres-Sánchez *et al.*, 2013). In the further combination of the UAVs with hyperspectral sensors, crop parameters can thus be recorded that cannot be seen with the unaided eye or that would require an unmanageable amount of work for human auditors. The combination of these sensors with existing technology enables site-specific weed control, using real-time kinematic global navigation satellite system (RTK-GNSS) controlled sprayers (Gerhards and Oebel, 2006; Mesas-Carrascosa *et al.*, 2016), but also expert system based decision support for crop production systems (Mortensen and Coble, 1991; Lindblom *et al.*, 2017). Beside site-specific weed management techniques, sensors were also used for the detection of herbicide-resistant weeds (Baker, 2008; Reddy *et al.*, 2014) and crop stress evaluation (Zhou *et al.*, 2016; Peteinatos *et al.*, 2016).

The state of the art of sensor technology for weed mapping is reviewed in Fernandez-Quintanilla *et al.* (2018). The more specific UAV-based aerial hyperspectral and red-green-blue (RGB) digital imagery for weed mapping has been demonstrated by Rasmussen *et al.* (2013) and Peña *et al.* (2013), but also in recent studies by Lambert *et al.* (2018) and de Castro *et al.* (2018). Although, considerable limitations in the ground sampling distance (GSD), compared to ground based approaches, UAVs are the preferable tool for site-specific weed management (Fernandez-Quintanilla *et al.*, 2018). The large number of research projects on the UAV weed detection for site-specific weed control shows not only the great potential, but also a wide variety of approaches. While many studies prefer the higher resolution RGB sensors (Rasmussen *et al.*, 2013), others argue for a combination of RGB and multispectral cameras (Peña *et al.*, 2013). Since multispectral cameras enable a more simple differentiation between vegetation and non-vegetation (e.g. soil, plant residues) pixels, they are often used as sensors for weed mapping despite their lower resolution compared to a RGB sensor. In addition to the vegetation detection, multispectral images also provide information about plant vitality (Zarco-Tejada *et al.*, 2013).

The use of aerial and ground based multi- or hyperspectral sensors for the vegetation stress assessments is based on the plant reflectance (Thenkabail *et al.*, 2018b). Various external and internal factors on the plant, may change its reflection of the incident light. These can be caused by genetic variability (Reddy *et al.*, 2014), but also by the deprivation of water or nutrients, by toxic substances or by temperature and daylight (Peteinatos *et al.*, 2016). Therefore, plant reflection depends not only on the cell tissue contents, but also on the efficiency of the photosystem and the associated changes of the absorbed light (Thenkabail *et al.*, 2018a). The stress origin often remains unclear if multiple factors affecting the plant vitality, as its manifestations in plant reflection overlap. This can be overcome in field experiments if assessments of a stressing factor can be compared to a untreated control variant. Therefore, these sensors could be employed for the evaluation of crop and weed responses to e.g. herbicide treatments.

The potential of sensors for herbicide sensitivity assessments by multi- or hyperspectral plant information (Reddy *et al.*, 2014; Lee *et al.*, 2014; Matzrafi *et al.*, 2017) or chlorophyll fluorometry (Kaiser *et al.*, 2013; Wang *et al.*, 2016) has already been demonstrated. The herbicide stress quantification using spectral data derived from imaging sensors or spectrometers can be assessed as other plant stress through changes in the plant reflectance values. The chlorophyll fluorometry is assessing the maximum quantum yield of the photosystem II, which can draw conclusion on the herbicide sensitivity of the investigated plant (Baker, 2008). By the evaluation of these parameters, information on the herbicide sensitivity or resistance can be derived, before visible symptoms of a herbicide injury occur (Reddy *et al.*, 2014; Kaiser *et al.*, 2013; Wang *et al.*, 2016). The focus of these studies was set on the investigation of possible parameters indicating a herbicide-resistance using chlorophyll fluorometry (Kaiser *et al.*, 2013), while Reddy *et al.* (2014) already trained a classifier on hyperspectral plant reflection of genetically homogeneous and heterogeneous weeds under artificial light conditions.

In order to extend the use of sensors for practical applications in plant protection, several aspects have to be overcome. The implementation of UAV weed mapping for site-specific weed control could drastically reduce the total amount of herbicide use if restrictions on data volume and ground sampling distance could be reduced. This would accelerate the computation time for the calculation herbicide application maps and therefore enable site-specific herbicide applications. A wider use of multispectral UAV imagery and ground-based hyperspectral plant information could significantly improve the quality of field experiment results if the auditor dependant visual estimates are supplemented by sensor derived datasets. The same applies to sensors used for herbicide-resistance detection in weeds. Here, portable field sensors would allow in-field sampling. A subsequent sensor data classification of a potential herbicide-resistant weed using pre-trained classifiers would enable direct in-season weed control of herbicide-resistant weeds prior to seed ripening.

1.1 Objectives

The further development of UAV and ground-based sensor information technologies for agricultural use cases to reduce the chemical burden to farmland, evaluate crop responses after following herbicide treatments and decision-support systems for herbicide-resistance management were investigated in this work, driven by the following objectives:

- the development and validation of site-specific herbicide applications, based on aerial imagery derived weed prescription maps

- the examination of hyperspectral aerial imagery to determine crop vitality in field trials to assist and expand the assessment of weed control strategies

- the examination of ground-based hyperspectral and chloropyll fluorometry sensor data to enable in-field and in-season detection of herbicide-resistant plants using classifiers.

1.2 Structure of the Dissertation

This dissertation is organised into 3 chapters covering the development, application and evaluation of UAVs and sensors for use in the management of weed control strategies. Chapter 1 provides an 'General Introduction' to the field of research and the specific research work of this dissertation, as well as the Structure of the Dissertation.

In chapter 2 the four individual research articles of this work are presented. The articles are arranged according to the course of weed control measures. Prior to the measure, the weed infestation on the area to be treated is mapped and a following site-specific herbicide application is presented in section 2.1. Subsequently, the effects of the treatment are evaluated regarding crop vitality and yield performance, which is described in the research article of section 2.2. Finally, it remains to be investigated whether there are resistant individuals among the possibly surviving weeds, which indicate the emergence of a herbicide-resistant population. Two systems to identify these individuals are presented in sections 2.3 and 2.4.

Finally a critical overview summarises the individual research articles in chapter 3. This 'General Discussion' provides additional insights, which exceed the contents discussed in the main articles.

Apart from the journal articles, contributions to national and international scientific conferences were part of this thesis. This work was supplementary to the studies presented and therefore not included in the current thesis:

- Menegat, A., Sievernich, B., Linn, A., Mink, R. and Gerhards R. Development of a standardised test system for detection of resistance against pre-emergence herbicides in *Proceedings 7th International Weed Science Congress*. "Weed science and management to feed the planet", Prague, Czech Republic, 2016. ISBN: 978-80-213-2648-4

- Mink, R. and Gerhards, R. Bewertung der Biomasseproduktion und Bodenbedeckung verschiedener Zwischenfrüchte durch UAV-Bildanalyse *in Programm Deutsche Phytomedizinische Gesellschaft Arbeitskreis-Herbologie, Nicht-Chemische bzw. Präzisions-Unkraut-Bekämpfungsverfahren*. Bingen, 2017

- Dutta, A., Gitahi, J., Ghimire, P., Mink, R., Peteinatos, G., Engels, J.-J., Hahn, M. and Gerhards, R.: Weed Detection in Close-range Imagery of Agricultural Fields using Neural Networks *in Beiträge Photogrammetrie-Fernerkundung-Geoinformatik-Kartographie-2018*. München, Germany, 2018. ISSN: 0942-2870

- Mink, R., Dutta, A., Peteinatos, G., Sökefeld, M., Engels, J.-J., Mozer, A., Mayr, W., Hahn, M. and Gerhards, R. Assessment of black-grass (*Alopecurus myosuroides* Huds.) densities using airborne imagery *in Book of Abstracts of the 18th European Weed Research Society Symposium*. Ljubljana, Slovenia, 2018. ISBN: 9789616998215.

- Linn, A., Mink, R., Peteinatos, G. and Gerhards, R. Herbicide efficacy estimation of ALS-inhibitors in *Stellaria media* L. and *Papaver rhoeas* L. *in Book of Abstracts of the 18th European Weed Research Society Symposium* Ljubljana, Slovenia, 2018. ISBN: 9789616998215.

- Mink, R., Linn, A., Gerhards, R. and Peteinatos, G. Classification of Herbicide Resistant *Papaver rhoeas* L. and *Stellaria media* L. Using Spectral Data *in Book of abstracts of the European Conference on Agricultural Engineering AgEng2018*. Wageningen, the Netherlands 2018. https://doi.org/10.18174/471678.

- Peteinatos, G. Mink, R., Linn, A. and Gerhards, R. Separating between herbicide sensitive and resistant *Papaver rhoeas* L. and *Stellaria media* L. Vill. plants, *in Sustainable Integrated Weed Management and Herbicide Tolerant Varieties*. American Farm School, Workshop, Thessaloniki, Greece, 2019.

Chapter 2

Research Articles

2.1 1st Study: Multi-Temporal Site-Specific Weed Control of *Cirsium arvense* (L.) Scop. and *Rumex crispus* L. in Maize and Sugar Beet Using Unmanned Aerial Vehicle Based Mapping

Article

Published in Agriculture 2018, 8(5), 65;

Received: 23 March 2018 / Accepted: 25 April 2018 / Published: 29 April 2018

https://doi.org/10.3390/agriculture8050065

Multi-Temporal Site-Specific Weed Control of *Cirsium arvense* (L.) Scop. and *Rumex crispus* L. in Maize and Sugar Beet Using Unmanned Aerial Vehicle Based Mapping

by Robin Mink[1], Avishek Dutta[2], Gerassimos G. Peteinatos[1], Markus Sökefeld[1], Johannes Joachim Engels[2], Michael Hahn[2] and Roland Gerhards[1]

1 Department of Weed Science, Institute of Phytomedicine, University of Hohenheim, 70599 Stuttgart, Germany

2 Faculty Geomatics, Computer Science and Mathematics, University of Applied Sciences Stuttgart, 70174 Stuttgart, Germany

Abstract

Sensor-based weed mapping in arable fields is a key element for site-specific herbicide management strategies. In this study, we investigated the generation of application maps based on Unmanned Aerial Vehicle imagery and present a site-specific herbicide application using those maps. Field trials for site-specific herbicide applications and multi-temporal image flights were carried out in maize (*Zea mays* L.) and sugar beet (*Beta vulgaris* L.) in southern Germany. Real-time kinematic Global Positioning System precision planting information provided the input for determining plant rows in the geocoded aerial images. Vegetation indices combined with generated plant height data were used to detect the patches containing creeping thistle (*Cirsium arvense* (L.) Scop.) and curled dock (*Rumex crispus* L.). The computed weed maps showed the presence or absence of the aforementioned weeds on the fields, clustered to $9\,m \times 9\,m$ grid cells. The precision of the correct classification varied from 96 % in maize to 80 % in the last sugar beet treatment. The computational underestimation of manual mapped *C. arvense* and *R. cripus* patches varied from 1 % to 10 % respectively. Overall, the developed algorithm performed well, identifying tall perennial weeds for the computation of large-scale herbicide application maps.

Keywords: digital elevation model; excessive green red vegetation index; patch spraying; site-specific weed control; UAV weed detection; weed mapping

2.1.1 Introduction

One of the major milestones in weed remote sensing technology research has been the implementation of Unmanned Aerial Vehicles (UAVs) as sensor carriers. Rasmussen *et al.* (2013) presented an estimation of plant soil cover from small and inexpensive aircraft systems evaluating the efficacy of mechanical weed harrowing in barley (*Hordeum vulgare* L.) and chemical weed control in oilseed rape (*Brassica napus* L. subsp. *napus*). Early season site-specific weed management in sunflowers based on UAV imagery is described in Torres-Sánchez *et al.* (2013). Both authors conclude that UAVs are useful to map weed pressure for site-specific weed management. Several UAV imaging sensors (e.g., Red, Green and Blue (RGB) and multispectral cameras), spatial resolutions and data analysis algorithms (e.g., Object-Based Image Analysis (OBIA)) are discussed in the literature (Rasmussen *et al.*, 2013; Torres-Sánchez *et al.*, 2013; Borra-Serrano *et al.*, 2015; Pérez-Ortiz *et al.*, 2016; Hung *et al.*, 2014; López-Granados *et al.*, 2016b,a; de Castro *et al.*, 2018; Fernandez-Quintanilla *et al.*, 2018).

Airborne imagery enables vegetation detection using vegetation indices, as reviewed in Salamí *et al.* (2014). Based on the Structure from Motion technique, 3D Models and Digital Elevation Models (DEM) can be created out of UAV imagery (Salamí *et al.*, 2014; Turner *et al.*, 2012). Combining this information can provide biomass estimations on barley (Tilly *et al.*, 2015) or yield predictions in maize, when combined with multispectral images (Geipel *et al.*, 2014). Further, weed identification has been performed by analysing vegetation height differences in a maize field using ground-based ultrasonic sensors (Andújar *et al.*, 2011a).

Christensen *et al.* (2014) discussed the complexity of various weed detection procedures, along with the generally low economic weed threshold levels (e.g. < 5 weeds m^{-2} or < 0.1 % weed cover). The authors remarked that the field area to be treated with herbicide highly depends on the used economic weed thresholds as also presented in Hamouz *et al.* (2018), Keller *et al.* (2014) and Longchamps *et al.* (2014). The information about local weed infestations becomes even more important in less competitive crops (e.g. sugar beet) or perennial weeds with a development stage dependent herbicide compatibility. Thus, high-resolution weed detection and multi-temporal mapping can play a major role in weed management.

Considering these possibilities and the well-known heterogeneous nature of weeds (Marshall, 1988; Johnson *et al.*, 1995), site-specific herbicide applications as reported by Gerhards *et al.* (2002) and Gerhards and Oebel (2006), should already be state of the art in today's farming practise. Yet, only a few site-specific herbicide application techniques have been commercially used (Christensen *et al.*,

2009). Fewer systems support the use of UAV weed mapping based on application maps.

It is crucial to fuse information from the different sensor systems and computing capabilities, used in modern precision farming, to crosslink all gathered field data for weed classification. Nevertheless, weed differentiation systems working in multiple crops by combining comprehensive field information have so far not been described in literature.

Consequently, the hypothesis was that the UAV mapping of herbicide-relevant weeds like creeping thistle (*Cirsium arvense* (L.) Scop.) can be accomplished by combining multiple sensor-based sources of precision farming information. Information on vegetation coverage and height derived from UAV imagery was connected to crop planting information (e.g. geo-coordinates of crop row locations and row space). By concatenating these data inputs, we propose a new methodology for improving the UAV weed mapping in arable fields. It is of particular interest to separate perennials like *C. arvense* and curly dock (*Rumex crispus* L.) from the rest of the fields' vegetation cover.

The objectives of the present study were (a) the development of an algorithm for the computation of herbicide application maps based on UAV weed mapping, (b) the subsequent use of these maps for weed spot spraying. (c) Realise the above objectives in row crops like maize at the three-leaf stage and in sugar beet between the cotyledon stage and the five-leaf stage. The computed application maps were transferred to a multiple tank spot sprayer for the site-specific treatment of *C. arvense* and *R. crispus*. This additional herbicide treatment was then applied along with a uniform herbicide application against annual grasses and broadleaf weed species.

2.1.2   Materials and Methods

Trial Sites and Precision Sowing

For the current study, four experiments were chosen at the Ihinger Hof research station (48.74°N, 8.93°E; 478 m a.s.l.) of the Hohenheim University, in southwest Germany during 2016 and 2017. The average 30-year annual temperature and precipitation are 8.4 °C and 738 mm. The fields were rotated within the two years of this study and they differed in their slope as shown in Table 2.1. The soil types differed from terra fusca, brown soil and pseudogley.

TABLE 2.1: Experimental details of the UAV weed mapping trials at Ihinger Hof, Germany.

Year	Crop	Trial Size (ha)	Terrain Slope (%)	Row Distance (cm)	Sowing Date	Seeds ha^{-1}
2016	sugar beet	2.07	2	50	21 April 2016	107,000
2016	maize	1.72	5	75	18 May 2016	94,000
2017	sugar beet	2.09	3	50	13 April 2017	107,000
2017	maize	2.07	2	75	11 May 2017	89,000

Sowing was realised using a Real-Time Kinematic (RTK)-corrected, Global Navigation Satellite System (GNSS)-assisted steering system on the tractor with a steady-state reference station located at the research station. The RTK-GPS enabled the pneumatic precision spaced planter (Maxima 2TI, Kuhn Maschinen-Vertrieb GmbH, Schopsdorf, Germany) to follow previously defined geographical coordinates for seeding with a precision of ≤ 2 cm.

Ground-Truth Weed Mapping

The weed plants were counted for each weed species at each centroid of a $9\,m \times 9\,m$ grid square, overlaid on the trial field. The totals were 173 grid squares in maize (2017) and sugar beet (2016) and 201 in maize (2016) and 220 in sugar beet (2017). Manual weed counting was performed one to three days before the site-specific weed control measurements. The collected weed scouting data of 0.4 m^2 per grid square was entered directly into a database, along with weed and crop cover and a photograph of the respective counting area. To this end weed species individuals were counted in a 0.1 m^2 counting frame with four repetitions around each grid square centroid, as tested and discussed in the literature (Gerhards *et al.*, 1996; Lamastus and Shaw, 2005; Gutjahr *et al.*, 2012).

These counting results were used to get a general overview about the field weed infestation. The database directly computed the plant density per m^2 for each species, using the single mapping repetitions per counting frame. The presence or absence of the later UAV-mapped *C. arvense* and *R. crispus* patches was additionally mapped outside the predefined counting points over the entire mapping area.

UAV and Sensor Setup

The weed mapping flights were performed using a hexacopter, model XR6 (geo-konzept, Adelschlag, Germany). The hexacopter was equipped with a Sony α 6000 (Sony Corporation, Minato, Japan) RGB camera and a Tetracam µMCA 4 Snap (Tetracam Inc., Chatsworth, CA, USA) multispectral camera including an electronic Incident Light Sensor (eILS).

The RGB image sensor lens focal length was 19 mm, with an image resolution of 6000×4000 pixels. Therefore, a Ground Sample Distance (GSD) of approximately 0.3 cm at a flight altitude of 15 m was achieved. The shutter was set to 1/1250 s and ISO between 200–1600 depending on actual lighting conditions prior to flight, resulting in an automatic aperture of $(f - x)$ with $x \in [7 - 10]$.

Multispectral images were taken at four narrowband spectral wavelengths of 670, 700, 740 and 780 nm using optical band-pass filters, each mounted in front of one of the camera's four image sensors. In addition to this, the uplooking eILS was mounted on the UAV. The eILS was capturing the downwelling radiation at the same wavelengths as those used in the aforementioned image sensors below. The camera was configured with a predefined focal length (9.6 mm), aperture (f/3.2), and automatic exposure time. The multispectral image sensor gathered 8-Bit, RAW file images with a resolution of 1280×1024 pixels, resulting in a GSD of approximately 1.51 cm at a flight altitude of 30 m. This flight altitude was chosen due to sensor limitations (e.g. image interval, battery power), to avoid the separation of flight plans in more than two parts.

Image triggering was software-controlled via a USB connection to the copter's flight-control-unit. At predefined points, image triggering was set as a flight plan action. Images were captured with a minimum overlap of 60 % in- and cross-track to enable photogrammetric calculations. The copter's flight altitude for use with the RGB sensor was set to 15 m and for the multispectral sensor to 30 m. If necessary, the flight plan was split into two parts for changing the two 24 V, 3700 mAh flight batteries. To overcome the decrease in the ground resolution, flight plans were adjusted to the ground morphology with a digital terrain model (DTM), using data from a previous flight. Therefore, the DTM was used to create the adjusted flight paths for the following flights.

Image Processing

The camera weed sampling flights were done at different crop growth stages, and flight dates, before and after the herbicide treatment. Weed mapping and application dates, along with the crop plant development stage, are given in Table 2.2. Prior to image alignment for the orthomosaic image creation, multispectral images had to pass through a pre-processing step. The alignment of the four single-channel images to the same field of view was carried out using the software PixelWrench2 (Version 1.2.3.1, Tetracam Inc., Chatsworth, CA, USA). In the same step, the images were corrected to the downwelling radiation, according to the data gathered by the eILS.

TABLE 2.2: Plant development stages at weed mapping and herbicide application dates.

Application Date	Herbicide Application [†]	Crop	Crop Development Stage	Mapping Date [‡]
20 May 2016	POST I	sugar beet	cotyledon stage	18 May 2016
10 June 2016	POST II	sugar beet	5 leaf stage	9 June 2016
22 June 2016	POST I	maize	3 leaf stage	20 June 2016
12 May 2017	POST I	sugar beet	cotyledon stage	11 May 2017
24 May 2017	POST II	sugar beet	4 leaf stage	23 May 2017
1 June 2017	POST I	maize	3 leaf stage	31 May 2017

† Herbicide applications were performed after crop emergence with specific herbicide doses per weed group as shown in Table 2.3 and Table 2.4 for each crop respectively.

‡ Mapping date includes UAV flight and manual weed mapping.

For generating georeferenced DEMs and orthomosaic images, the 3D reconstruction software PhotoScan Professional Edition (Version 1.3.4, Agisoft LLC, St. Petersburg, Russia) was used. Beside the airborne images, the copters flight information and positions of the RTK-GPS-referenced Ground Control Points (GCPs) located inside the trial and around the borders (Torres-Sánchez *et al.*, 2013; Geipel *et al.*, 2014) were passed to the software.

Generating Weed Control Maps Based on UAV Imagery

To compute large area site-specific herbicide application maps, a robust and relatively fast to compute algorithm was designed. The targeted output was an application map to provide the respective geo-coordinates and decisions for one tank of a three tank site-specific sprayer. For further processing, the multispectral and RGB orthomosaic images were loaded as raster objects in the statistical computation software R (Core Team R, 2018), using the packages raster, version: 2.6-7 (Hijmans *et al.*, 2017) and rgdal version: 1.2-18 (Bivand *et al.*, 2017).

The crop row locations were known from the RTK-GNSS-assisted seeding and therefore a predefined AB-line was available from the beginning of the cultivation. Based on this AB-line, all crop row positions were calculated and used to calculate a polygon layer. The crop rows were deleted out of the orthomosaic images by overlaying the crop row polygon layer. The width of the crop row polygon layer was adjusted, based on the actual crop growth stages.

For further discrimination of weeds and soil, the Normalised Difference Vegetation Index (NDVI) was calculated based on the multispectral images, adapted from Rouse *et al.* (1973) using the wavelengths 670 and 780 nm, as shown in Equation 2.1. The resulting NDVI raster object was used as a mask layer as proposed by Liebisch *et al.* (2015) and demonstrated in Roth and Streit (2017) on the relative RGB image. All parts of the raster object with an NDVI > 0.2 were considered as vegetation. If multispectral data were not available, weed maps were calculated only using the RGB images. The Excessive Green Red index (ExGR, Equation 2.2) was calculated out of the three layers of the processed RGB orthomosaic image (Ortho), representing the crop fields inter-row space. The ExGR combines two colour indices: the Excessive Red index (OrthoExR, Equation 2.3) as shown in Meyer *et al.* (1999) and the Excessive Green index (OrthoExG, Equation 2.4) as shown in Woebbecke *et al.* (1995). The result derived from the subtraction of the two aforementioned indices is shown in Equation 2.2. Thereby the background noise in the excessive green grayscale image was reduced and a zero threshold could be applied for the creation of a binary image. The ExGR binary orthomosaic image (OrthoExGR) was generated using the ExGR zero threshold method as introduced by Meyer *et al.* (2004) and suggested by Hamuda *et al.* (2016). Prior to the index calculation, all single layers had to be normalized by the max value of each RGB channel.

$$Normalized\ Difference\ Vegetation\ Index = \frac{780\ nm - 670\ nm}{780\ nm + 735\ nm} \qquad (2.1)$$

$$Ortho_{\mathsf{ExGR}} = Ortho_{\mathsf{ExG}} - Ortho_{\mathsf{ExR}} \qquad (2.2)$$

$$Ortho_{\mathsf{ExR}} = \frac{1.4 \times Ortho_{\mathsf{red}} - Ortho_{\mathsf{green}}}{Ortho_{\mathsf{red}} + Ortho_{\mathsf{green}}} \qquad (2.3)$$

$$Ortho_{\mathsf{ExR}} = \frac{2 \times Ortho_{\mathsf{green}} - Ortho_{\mathsf{red}} - Ortho_{\mathsf{blue}}}{Ortho_{\mathsf{green}} + Ortho_{\mathsf{red}} + Ortho_{\mathsf{blue}}} \qquad (2.4)$$

We considered that each pixel in the classified OrthoExGR raster object can be attributed to vegetation, or to secondary substances, like soil or organic matter, based on the result of the binary map, as shown in Equation 2.5. The classified pixels in vegetation or soil and organic matter had the geographic coordinates north (n) and east (e).

$$f : X_{(n,e)} = \begin{cases} 1, \forall \ Ortho_{ExGR^{(n,e)}} \in vegetation \\ 0, \forall \ Ortho_{ExGR^{(n,e)}} \notin vegetation \end{cases} \tag{2.5}$$

Since the classification was conducted with 0 as the threshold value, the OrthoExGR image was binarized as shown in Equation 2.6. The result is the classified Orthoimage of Equation 2.5.

$$f : X_{(n,e)} = \begin{cases} 1, \forall \ Ortho_{ExGR^{(n,e)}} \in (-\infty, 0] \\ 0, \forall \ Ortho_{ExGR^{(n,e)}} \in (0, +\infty) \end{cases} \tag{2.6}$$

Apart from the vegetation information, height information was also used for the discrimination algorithm. The separation of the inter-row vegetation into lower and taller weeds was carried out based on the Canopy Height Model (CHM). The CHM could be calculated by subtracting the Digital Terrain Model (DTM) raster object from the Digital Surface Model (DSM) raster object (Turner *et al.*, 2012; Geipel *et al.*, 2014) as shown in Equation 2.7. The CHM presented the height for every single pixel in the corresponding raster object.

$$CHM = DSM - DTM \tag{2.7}$$

The DTM was computed by UAV imagery gathered directly after seeding. For calculating the DSM, UAV images from the respective mapping dates were used. The general computation of this digital elevation model was performed as a prior step during the orthomosaic computation from the 2D images. The CHM threshold for separating lower and higher vegetation of the inter-row space was set to a height of 6 cm above ground as shown in Equation 2.8.

$$f : C_{(n,e)} = \begin{cases} 1, \forall \ CHM^{(n,e)} \in [0.06, +\infty) \\ 0, \forall \ CHM^{(n,e)} \in (0, 0.06) \end{cases} \tag{2.8}$$

All Vegetation pixels above the used threshold were regrouped in a new class of the raster layer. To verify CHM raster cells as vegetation, we overlaid the crop row excluding ExGR raster to the CHM raster (Equation 2.9).

$$f : WHM_{(n,e)} = X \bullet (X + C) \qquad (2.9)$$

The resulting Weed Height Model (WHM) raster object, therefore, combined field and seed planning precision framing data with models calculated from UAV based airborne images. It provided information about the field weed infestation in two classes and the extracted information was used in the site-specific herbicide sprayer on-board computer database.

Site-Specific Herbicide Application in Field Trials

Site-specific herbicide applications were realised with a trailed multiple-tank spot-sprayer, with three separate hydraulic systems. The sprayer was equipped with RTK-GPS and section control to adjust the application of different herbicides at specific field locations (Gerhards and Oebel, 2006; Gutjahr *et al.*, 2012). Using three separated hydraulic systems on the sprayer, three different herbicide mixtures were stored and applied wherever it was designated from the application maps. Each hydraulic circuit had its own tank and boom. The booms had a total width of 21 m and were divided into seven sections each of 3 m width.

The sprayer's section control turned on and off the sections used for the site-specific application via solenoid valves. The same terminal also regulated the herbicide doses for the two other hydraulic systems, applying herbicides on the complete field.

Herbicide application maps of sugar beet and maize trials were used to apply products as shown in Table 2.3 (sugar beet) and Table 2.4 (maize) for the mapped group of weeds respectively. The composition of the herbicide mixtures was the same in both study years. The herbicide selection for the post-emergence treatments of the mapped species was based on regional guidelines in Germany and is given in Tables 2.3 and 2.4. As shown in Table 2.2, herbicide applications, weed mapping and flight dates were synchronized for each treatment within a three-day period. Inside each field, two of the grid squares were established as an untreated control by manually adjusting the herbicide application maps prior to spraying.

TABLE 2.3: Weed-specific herbicides applied with a three-tank spot-sprayer in sugar beet trials.

Weed group	Herbicide a.i.	Product name	Formulation	Product concentration†	Herbicide rate g a.i. ha^{-1}†
Dicotyledon	*phenmedipham*	Betasana SC®	160 g L^{-1}	0.52 L 100 L^{-1}	208.0
	ethofumesat	Ethosat 500®	250 g L^{-1}	0.18 L 100 L^{-1}	112.0
	metamitron	Goltix Gold®	700 g L^{-1}	0.72 L 100 L^{-1}	1254.5
Cirsium arvense *Rumex crispus*	*clopyralid*‡	Lontrel 720 SG®	720 g L^{-1}	32 g 100 L^{-1}	79.9
Monocotyledon	*fluazifop-P-butyl*	Fusilade max®	125 g L^{-1}	0.16 L 100 L^{-1}	50.8

† All herbicides applied at 250 L ha^{-1}.

‡ *clopyralid* was applied with 0.5 L ha^{-1} liquid paraffin.

TABLE 2.4: Weed-specific herbicides applied with a three-tank spot-sprayer in maize trials.

Weed group	Herbicide a.i.	Product name	Formulation	Product concentration†	Herbicide rate g a.i. ha^{-1}†
Dicotyledon	*bromoxynil*	Bromotril®	225 g L^{-1}	0.17 L 100 L^{-1}	112.5
	mesomitrone	Callisto®	100 g L^{-1}	0.23 L 100 L^{-1}	66.0
Monocotyledon	*foramsulfuron*	Maister®	30 g L^{-1}	0.20 L 100 L^{-1}	17.7
	+ *iodosulfuron-methyl-natrium*		1 g L^{-1}		0.6
	+ *isoxadifen-ethyl* (safener)		30 g L^{-1}		17.7
Cirsium arvense	*clopyralid*	Efigo®	267 g L^{-1}	0.08 L 100 L^{-1}	59.3
	+ *picloram*		67 g L^{-1}		14.9

† All herbicides applied at 290 L ha^{-1}.

In 2016, a hexacopter with an RGB and a four channel multispectral camera was used for weed mapping. The GSD of the gathered images during these flights was insufficient for computing valuable weed maps when following the presented protocol. Therefore, site-specific weed control application was completely based on manual weed maps in 2016. The complete workflow, including the different computation steps is presented in the flowchart of Figure 2.1. The procedure described therein was applied to every XR6 flight dates without changes in the main algorithm.

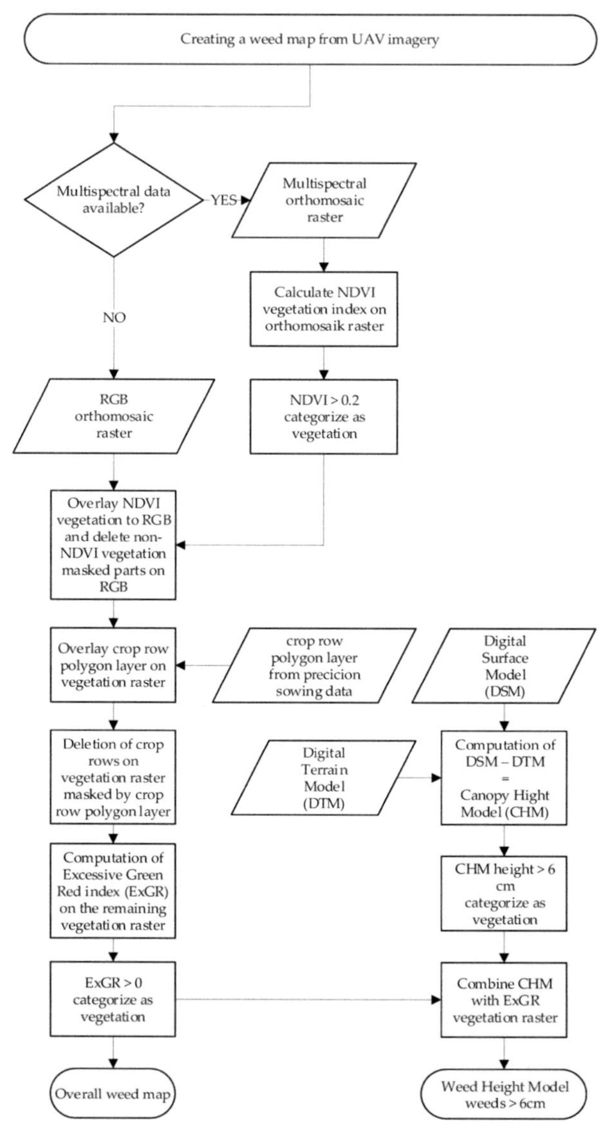

FIGURE 2.1: Data input and processing steps to compute UAV imagery-based weed maps for site-specific herbicide applications in a patch sprayer.

2.1.3 Results

The automatic flight altitude adjustment by following a digital elevation model
resulted in a steady GSD of the orthomosaic images and DEMs. A binary raster
was created using the workflow as described in Figure 2.1 on RGB and multispectral
images gathered by an UAV. Figure 2.2 shows the different processing stages of the
binary creation. The raster containing the height information is shown along with the
raster containing the vegetation index and the crop row polygons, all overlaid on the
starting RGB raster, showing a *C. arvense* patch in sugar beet at the 4 leaves crop
development.

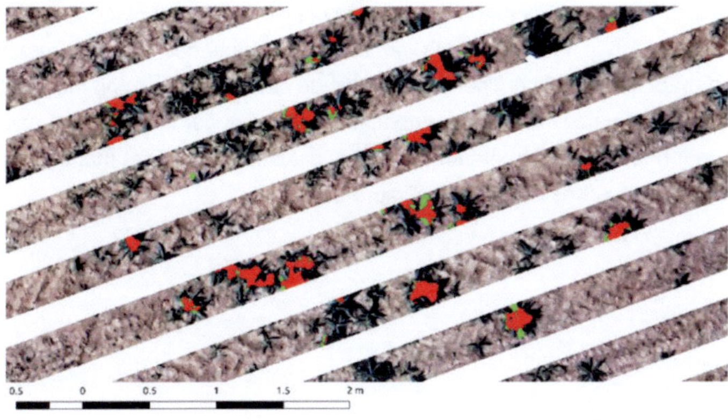

FIGURE 2.2: Overlay of multiple raster layers onto the original RGB
image of a *C. arvense* patch in the sugar beet field previous to the
POST II herbicide application at four leaf crop development stages.
The crop row polygons are depicted as white stripes. Red parts
indicate ExGR vegetation index approved plant parts of the CHM
> 6 cm. Green pixels show non-ExGR approved parts of the CHM
> 6 cm.

Although, overlaying different raster files from different flights, e.g. DTM
and DSM for CHM calculation worked as expected, the image resolution slightly
decreased during the, resampling of the thereby adjusted raster files.

In this study, the weed species distribution was heterogeneous within the
experiments. In the 2017 maize trial the main weed species was corn sowthistle
(*Sonchus arvensis* L.) with a mean plant density of 15 plants m^{-2} and bearbind
(*Fallopia convolvulus* (L.) Á. Löve) with a mean plant density of 12 plants m^{-2}. In the
maize trial of 2016 it was common lambsquarters (*Chenopodium album* L.) with 55

plants m^{-2} and annual bluegrass (*Poa annua* L.) with 24 plants m^{-2}. In some distinct places of the maize trial in 2016 the weed density was very high, with up to 375 *P. annua* and 150 *C. album* plants m^{-2}. In the sugar beet trials the weed density also changed between the application dates. While the first herbicide application of 2017 had to treat weed densities of up to 450 black-grass (*Alopecurus myosuroides* Huds.) individuals, the number decreased to 280 plants m^{-2} at the second treatment. Another major weed in all trials was chamomile (*Matricaria chamomilla* L.) with plant densities between seven plants m^{-2} in the sugar beet trial of 2016 and 108 plants m^{-2} in the sugar beet trial 2017. In general the weed species composition was linked closer with the trial field, than with the planted crop. In 2017 patches containing *C. arvense* or R. *crispus* were manually mapped only in 16 distinct places in maize. On the other hand, up to 114 grid points with the abovementioned weeds were found in the sugar beet field of 2017. Therefore, herbicide savings concerning these patches was 90 % in the maize POST I treatment, 45 % in the sugar beet POST I and 43 % in the sugar beet POST II treatment. The spatial distribution of the manual and UAV classified weed patches are given in Figure 2.3.

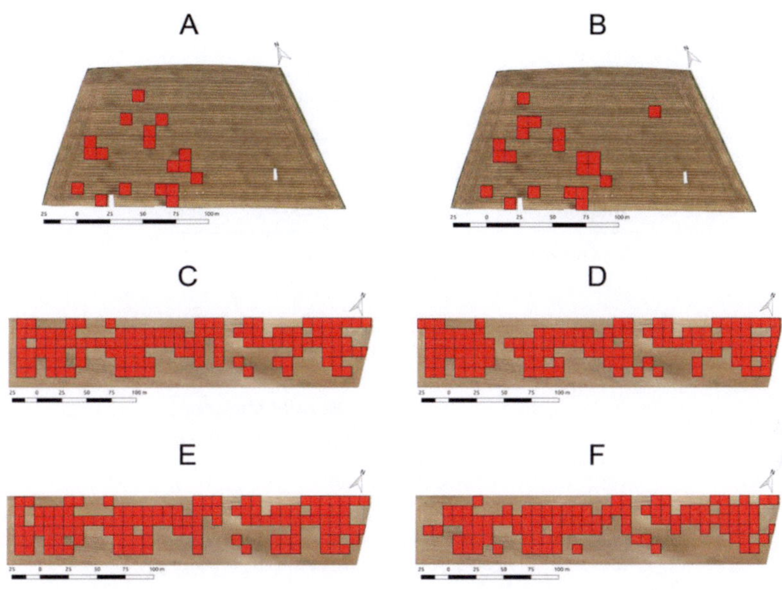

FIGURE 2.3: Maps identifying manual (**A,C,E**) and UAV (**B,D,E**) mapped *C. arvense* and *R. crispus* from the vegetation period of 2017. Red squares indicate the presence of ≥ 1 *C. arvense* or *R. crispus* plant. The orthoimages and application maps were computed prior to the site-specific herbicide treatments in maize POST I (**A,B**), sugar beet POST I (**C,D**) and sugar beet POST II (**E,F**).

Weed threshold levels for herbicide application against all mapped weeds were set at 0 plants m^{-2} in sugar beet and maize. Therefore, site-specific treatment of the UAV classified weeds (*C. arvense* and *R. crispus*) was only possible using a GPS-controlled patch sprayer with multiple separated hydraulic circuits. Using this sprayer in the field, herbicide application against monocotyledon and dicotyledon weeds was realised using two of the three hydraulic systems of the sprayer, while the site-specific treatment of the UAV mapped weeds (*C. arvense* and *R. crispus*) was applied with the third hydraulic system.

Compared with the manual weed maps, the UAV categorized maps showed an accuracy of 96 % in the maize POST I treatment, 90 % in the sugar beet POST I and 80 % in the POST II treatment, compared with the manual weed maps. The overall manual and UAV weed patch identification is presented in confusion matrices in Tables 2.5, Table 2.6 and Table 2.7.

TABLE 2.5: UAV and manual mapping prior to maize POST I treatment.

		Manual Mapping	
		Weed-free	Weed
UAV	Weed-free	16	2
mapping	Weed	5	150

Weed refers to *C. arvense* and *R. crispus* > 6 cm, n = 173.

TABLE 2.6: UAV and manual mapping prior to sugar beet POST I treatment.

		Manual Mapping	
		Weed-free	Weed
UAV	Weed-free	114	7
mapping	Weed	14	85

Weed refers to *C. arvense* and *R. crispus* > 6 cm, n = 220.

TABLE 2.7: UAV and manual mapping prior to sugar beet POST II treatment.

		Manual Mapping	
		Weed-free	Weed
UAV	Weed-free	103	23
mapping	Weed	21	73

Weed refers to *C. arvense* and *R. crispus* > 6 cm, n = 220.

Compared to the respective manual weed map, the UAV identification overestimated weed patches of 3 % in maize POST I and 10 % in sugar beet POST II, shown in Tables 2.5, Table 2.6 and Table 2.7. However, the number of UAV underestimated patches was 50 % lower. The only exception to this statement was the sugar beet POST II treatment.

2.1.4 Discussion

The large-scale precision farming field experiments conducted in this study proved that the herbicide output can be reduced, when combining UAV imagery and plant seeding information, for computation of detailed application maps.

The presented workflow could be used in maize and sugar beet with generally low weed threshold levels (Gerhards and Oebel, 2006; Williams II *et al.*, 2000). Thus, the sprayer was only turned off at locations where none of the respective weeds was found. For weed recognition, similar results were found using tractor mounted, camera based, online systems (Gerhards *et al.*, 2002; Gerhards and Oebel, 2006; Tian *et al.*, 1999). Although all these studies integrated camera mapped weeds in the site-specific management, the results in the present study clearly stated the possibility of weed categorization and herbicide savings, even though the UAV data were only computed for two weed species.

While the spatial resolution of the UAV images for weed recognition was lower compared to tractor-mounted cameras, the UAV mapping has its advantages. The UAV weed mapping can be conducted, before the application and without soil compaction. Therefore, farmers can calculate the amount of herbicide needed on the field beforehand. Also, a multi-temporal, weekly surveillance can be realised for surveying actual weed infestation in order to meet the economic threshold levels for herbicide applications.

Using polygons to exclude crop rows at a known position speeds up the data analysis and reduces the total data size that not only needs to be computed, but also transferred, and stored. Although the intra-row space was not yet investigated, there was no considerable loss of data concerning the investigated weed patches due to the intra-row deletion. The generally equivalent weed coverage between inter row and crop row areas was also stated in Longchamps *et al.* (2012), although they found a partially higher weed infestation in the crop rows. Furthermore, the total size of the dataset was reduced by up to 30 % before entering the main algorithm computation. Extracting plants from the RGB orthomosaic images with the ExGR vegetation index and the use of a steady zero threshold level for binary image calculation was also possible for the UAV imagery. This complies with the proposal of Meyer *et al.* (2004); Hamuda *et al.* (2016) The fixed threshold level for separating vegetation from soil is a key element towards automation of image-based weed recognition. Especially the differences in soil and vegetation appearance, e.g., wet soil, dense vegetation or the changing light conditions between flights, can cause deviations from predefined thresholds. Thus, the combination of the ExG and ExR vegetation indices in the ExGR provides an option to circumvent this problem.

Even though this robust and fast computable vegetation index for *C. arvense* and *R. crispus* worked, the loss of information in our data set did not allow the analysis of small broad-leaved and grass weeds. To realise the separation of smaller weeds at the cotyledon stage, for example by using the OBIA algorithm, higher resolution images will be needed. If this can be achieved, UAV weed mapping of different species and herbicide treatments at early weed plant stages, as needed in sugar beets, can be realised. This well-known coherence between image sensor resolution, flight altitude and vegetation index thresholds is also discussed by López-Granados *et al.* (2016b), Mesas-Carrascosa *et al.* (2015) and Mesas-Carrascosa *et al.* (2016). Again, the need of high resolution image sensors becomes apparent. This was also shown in our multispectral data sets providing an almost four times lower GSD. The NDVI orthoimages that we calculated from our multispectral images could be used for pre-categorisation of our RGB orthoimages. However, single weed detection was not possible. On the other hand, the multispectral images could be corrected to the downwelling sunlight using the eILS data. Therefore, the result of the NDVI >0.2 (Liebisch *et al.*, 2015) was relatively consistent.

Combining information by overlaying the RGB calculated ExGR vegetation index and CHM already provided information about the field weed infestation. It enabled the separation of single weeds in different herbicide application relevant categories. Compared to simple colour-based algorithms, a secondary information source was added. A comparable approach is also described by de Castro *et al.* (2018) for weed detection in common sunflower (*Helianthus annuus* L.) and cotton (*Gossypium* spp.). With this additional pixel height information, plants can not only be identified by their colour but also by their height. In our case, this provided already enough information for separating *C. arvense* and *R. crispus* from other crop and weed plants, found in the crops inter-row space.

The detection of *C. arvense* plants at 6 cm enables the control at the plant's compensationpoint, where the acropetal allocation of carbohydrates turns into a basipetal movement of photosynthates (Nkurunziza and Streibig, 2011). At this point the plant is highly susceptible to control measures, since the root system has the lowest regenerative capacity (Welton *et al.*, 1929). The typical appearance in patches of C. arvense, which is caused by its mainly vegetative reproduction inside arable fields (Håkansson, 2003), can also provide information of patches appearing in the near future or the next vegetation period.

From a practical point of view, it is better to overestimate the presence of *C. arvense*, than underestimate it. Even in this case, herbicide savings can be achieved, especially at low *C. arvense* infestations. Using this raster layer for OBIA analysis is not yet compatible, since no distinct shape structures of the weeds are available anymore, as shown in Figure 2.2.

The generation of orthomosaic images and CHMs with a steady spatial resolution is essential to follow the weed mapping workflow presented in this paper. The

DTM following flight plan was a useful adjustment in our fields to follow the slope of up to 10 m. The unintended increase of the flight altitude caused by the terrain would have caused a loss in the spatial resolution. When flying the UAV at low altitudes of 15 m, a slope of 5 m, as tested in our study fields, would reduce the spatial resolution in the lowest parts of the field from a GSD of 0.3 cm to 0.4 cm, if the flight altitude is not adjusted. Furthermore, the precise overlaying of rasters was also supported by georeferencing the images using the RTK-GPS corrected GCPs, which helped to improve the accuracy of the coordinate system on the orthoimages.

2.1.5 Conclusions

(a) We can conclude that the presented aerial imagery computation procedure is able to identify the presence or absence of weeds above a defined height threshold level in row crops. The use of a field CHM in combination with the Vegetation Index ExGR and crop row geo-coordinates enabled the separation of *C. arvense* and *R. crispus* from the rest of the vegetation by their height with an overall accuracy of 96 % in the maize POST I, 90 % in the sugar beet POST I and 80 % in the sugar beet POST II treatments. (b) The output data enables the control of site-specific weed management strategies Resulting in herbicide savings of 90 % in the maize POST I treatment and 43 % in the sugar beet POST II treatment. (c) The same algorithm was able to work in two row crops. Thereby the aggregation of different precision farming data sources lead also to a reduction in data size and needed computation power, already in early processing steps. This can be a key element in agricultural fields where fast Internet connections for transferring data to high-performance computing clouds are not yet guaranteed. In order to mainstream the processing chain into a simple fly and treat application, further investigations are needed, specifically orthomosaic image georeferencing and overlaying of multi-temporal UAV orthoimage rasters, which are labour-intensive.

Author Contributions

R.M. and M.S. were responsible for the field experiments and data collection. R.M. and G.G.P. developed the main workflow and computing algorithm. R.M., A.D., G.G.P. and J.J.E. wrote and revised the manuscript. M.H. and R.G. supervised the experiment and revised the manuscript.

Funding

This project was supported by funds from the Federal Ministry of Food and Agriculture (BMEL) based on a decision of the Parliament of the Federal Republic of Germany via the Federal Office for Agriculture and Food (BLE) under the innovation support programme.

Acknowledgments

The authors would like to acknowledge the great work of the student Fernando Laureano Palacios Duarte.

Conflicts of Interest

The authors declare no conflict of interest.

2.2 2nd Study: Sensor-based Evaluation of Maize (*Zea mays*) and Weed Response to Postemergence Herbicide Applications of *Isoxaflutole* and *Cyprosulfamide* Applied as Crop Seed Treatment or Herbicide Mixing Partner

Article

Published in Pest Management Science; doi:10.1002/ps.5715

Received: 16 September 2019 / Accepted: 08 December 2019 / Published: 11 December 2019

https://doi.org/10.1002/ps.5715

Sensor-based Evaluation of Maize (*Zea mays*) and Weed Response to Postemergence Herbicide Applications of *Isoxaflutole* and *Cyprosulfamide* Applied as Crop Seed Treatment or Herbicide Mixing Partner.

Running title: *Cyprosulfamide* seed treatments

by Robin Mink, Alexander Ingo Linn, Hans-Joachim Santel and Roland Gerhards

Department of Weed Science, Institute of Phytomedicine, University of Hohenheim, 70599 Stuttgart, Germany

Abstract

Some maize post-emergence herbicides obtain their crop/weed selectivity only through the use of chemical crop safeners. Safeners improve the tolerance of maize to herbicidal active ingredients. In order to investigate the crop response to (*cyprosulfamide*) spray application and seed treatment, greenhouse and field trials were conducted on three maize development stages (2-, 4-, and 6-leaf stage). Visual estimations on crop vitality were compared to ground-based and airborne hyperspectral and multispectral sensors. The reduction of *cyprosulfamide* by 88% when applied as seed treatment did not significantly reduce maize biomass yields at the field. Additionally, the safener seed treatment leads to significantly lower crop stress at the 4-leaf development stage at the field trial. The sensor-based and visual estimated crop deterioration in both trials was stronger in the *cyprosulfamide* seed treatments compared to the spray applications but was found to be transient in the field trial. The hyperspectral sensor and multispectral camera data was comparable with R^2 = 0.84 (CropSpec vegetation index) and R^2 = 0.64 (Green Normalised Difference Vegetation Index). The sensor-based results were consistent with the visual damage assessment and indicated significant differences between treatments already at one day after treatment.

Keywords: crop stress measurement, remote sensing, spectrometer, UAV multispectral imagery, sensor fusion, spectral indices, weed management

2.2.1 Introduction

Weed control with herbicides and safeners evolved into an important management strategy in the global maize (*Zea mays* L.) cultivation. As a result of the constantly increasing herbicide use, social and environmental concerns on the chemical burden imposed on the environment meanwhile take center stage and trigger political action. On the other hand, effective weed control, until at least the 4-leaf stage of maize, is indispensable to ensure yield stability (Hall *et al.*, 1992).

Yet, many of the chemicals available as post-emergence herbicides for the use in maize lack a sufficient selectivity needed for the required level of weed control without harming the maize plants (Kraehmer *et al.*, 2014b). While the selectivity of some herbicidal compounds is based on differences in the foliage morphology, differential compartmentation or inactivation by metabolism between the crop and the weeds, others need additional support to become selective (Hock *et al.*, 1995; Dan Hess, 2018). Herbicide-tolerant transgenic organisms and the use of safeners represent two additional methods to ensure a satisfying herbicide selectivity (Owen and Zelaya, 2005; Johnson *et al.*, 2000).

Safeners are chemical compounds acting as a selective herbicide antidote on monocotyledonous crop plants. Thereby the safener treatment must not reduce the herbicide activity in the target weed species (Davies and Caseley, 1999). However, in rare cases, the application of the safener as a mixing partner of the herbicide active ingredient can lead to the unintentional protection of the monocotyledonous weeds to be controlled (Kraehmer *et al.*, 2014a) and may increase the necessary herbicide dose and thus the chemical burden on the fields.

Site-directed application of safeners is possible when removed from the coformulation and applied directly to the seed as a seed treatment (Kraehmer *et al.*, 2014a; Hatzios and Hoagland, 1989). Applied in this way, there is no need for the safener to be active in the crop only. Nevertheless, today's safeners are mostly sold as crop-selective ready-to-use mixtures with the herbicidal compounds. Thus, the supplier retains control of the herbicide partner and unregistered uses can be avoided.

Isoxaflutole (IFT, herbicide) and *cyprosulfamide* (CSA, safener) represents one of these herbicide/safener pairs. IFT belongs to the *4-hydroxyphenylpyruvate dioxygenase* (HPPD) inhibitors, while CSA safens *acetolactate synthase* (ALS) inhibitors (e.g. *thiencarbazone-methyl*) and HPPD-inhibitors in maize (Santel, 2012; Robinson *et al.*, 2013).

To investigate the effects of CSA on IFT-treated maize, laboratory studies,(Pallett *et al.*, 1998) as well as greenhouse (Zhang *et al.*, 2007) and field trials have been carried out (Robinson *et al.*, 2013; Knezevic *et al.*, 1998). Usually, destructive measurements or visual evaluations of plant vigour parameters are carried out for the evaluations in greenhouse and field trials. Although both methods are common practice, they have considerable disadvantages. Visual estimations on plant health status are always affected by an auditor-dependent subjective variation (Andújar *et al.*, 2011b), while destructive measurements may influence the course of the trial and its results. These adverse measurement methods can be substituted or reduced by employing an additional non-destructive sensor derived information. Here, unmanned aerial vehicles (UAV) equipped with multispectral sensors can provide highly detailed crop stands and vigour information (Zaman-Allah *et al.*, 2015), independent of orbit time or cloud cover. By measuring information on plant stress can be derived (Lichtenhaler *et al.*, 1998).

Depending on the experimental question, different vegetation indices (VIs) are suitable for a precise evaluation of the plant parameters measured. Likewise, VIs have already been used on IFT applications to estimate changes in the pigment ratio in maize plants following an IFT application(Yancheva *et al.*, 2016) and correlated well with the chlorophyll content of leaves (Lichtenhaler *et al.*, 1998).

Based on these findings we investigated the following hypotheses: VIs, derived from multispectral imagery and hyperspectral sensor information can provide detailed information about the crop vitality and development after the IFT application on CSA treated maize plants. By comparing the sensor derived data inputs with destructive measurements and auditor estimations we discussed the use of the ground-based hyperspectral sensor and multispectral imagery in a greenhouse and a subsequent field trial.

The objectives of the present study were (a) to identify if the required safening potential of CSA in maize can be achieved through a CSA seed treatment and at the same time reduction of the safener rate per hectare, and (b) if the crop response to the CSA seed treatment can be measured by sensor data after IFT application.

2.2.2 Materials and Methods

Experimental Design

The crop damage reduction potential of CSA to IFT post-emergence applications was examined in a greenhouse trial and a following field trial with maize (Cultivar DAVOS, Deutsche Saatveredelung AG, Lippstadt, Germany). CSA was sprayed as a premix formulation with IFT or as a CSA seed treatment with the subsequent IFT spraying application at three consecutive crop development stages (Table 2.8).

2.2. 2nd Study: Sensor-based Evaluation of Maize (Zea mays) and Weed Response to Postemergence Herbicide Applications of Isoxaflutole and Cyprosulfamide Applied as Crop Seed Treatment or Herbicide Mixing Partner

TABLE 2.8: Herbicide and safener treatments of the greenhouse pot trial and the field trial. The combined safener and herbicide application (COMB) and the separated application with a safener seed treatment (SPLIT) were tested at three consecutive maize development stages.

Treatment	Maize leaf stage	Safener application method[†]	Product name	Product concentration g L^{-1}	isoxaflutole g ha^{-1}[‡]	cyprosulfamide application rate	Application date field
COMB-2	2	spray	Balance[®] Flexx	240	150	150 g ha^{-1}	21.05.2018
COMB-4	4	spray	Balance[®] Flexx	240	150	150 g ha^{-1}	28.05.2018
COMB-6	6	spray	Balance[®] Flexx	240	150	150 g ha^{-1}	04.06.2018
SPLIT-2	2	seed treatment	Balance[®] Pro	400	150	0.2 mg seed^{-1}	21.05.2018
SPLIT-4	4	seed treatment	Balance[®] Pro	400	150	0.2 mg seed^{-1}	28.05.2018
SPLIT-6	6	seed treatment	Balance[®] Pro	400	150	0.2 mg seed^{-1}	04.06.2018
Untreated control	2; 4; 6	-	-	-	-	-	-
Herbicide control[§]	4	-	Balance[®] Pro	400	150	-	28.05.2018

† Safener treated seeds, as well as the seeds of the untreated and manual control are additionally treated with 40 μL Peridiam[®] Evolution EV302 (Bayer CropScience AG, Monheim am Rhein, Germany) and 160 μL deionised water.

‡ Herbicide was applied at 200 L ha^{-1}

§ Treatment was only carried out in the field trial

Seed Treatment: The safener seed treatment included 40 μL Peridiam[®] Evolution EV302 (Bayer CropScience AG, Monheim am Rhein, Germany) as a color marker, 40 μL CSA (500FS, β = 500g L^{-1}) and 160 μL deionised water. The untreated control seed was treated the same way but without CSA. Germination tests were carried out, according to International Seed Testing Association (2019) at the sowing dates of the respective trials.

Prior to the seed treatment twofold repeated greenhouse pretests with four different CSA dosages (0.1, 0.2, 0.3 and 0.4 mg seed^{-1}) as seed treatments and an untreated control were carried out. The IFT applications followed directly as pre-emergence application one day after sowing. IFT was applied at the dosages 50, 100 and 200 g ha^{-1} at 200 L ha^{-1}. Additionally, a germination test was conducted at two temperature regimes of 10 $^\circ$C and 20 $^\circ$C at the four CSA seed treatment rates.

Greenhouse Pot Trials: The experiments were carried out on maize plants in the greenhouse of the University of Hohenheim (48.71°N, 9.21°E; altitude 370 m a.s.l.), in 2017 and 2018. Each experiment was repeated in time and set up as a two-factorial pot trial in a randomized complete block design with three replications. The pots (11 × 11 × 12 cm^3) were filled with soil containing 50 % loam, 25 % sand and 25 % compost. In each pot, one maize seed was placed at a depth of 3 cm. The herbicide and safener applications were carried out in a precision laboratory track sprayer (Detron AG, Stein, Swiss) using a flat fan nozzle (8002EVS, TeeJet[®], Spraying Systems Co., Wheaton, IL, USA), simulating a carrier volume of 200 L ha^{-1}(Table 2.8).   The plants were daily watered from above until the end of the trial. The greenhouse temperature regime was regulated by controlled outside air supply, preventing temperatures lower than 15 $^\circ$C and following the daily temperature fluctuations. No additional greenhouse lights were used.

Field Trial: A field trial was carried out on maize plants at the Ihinger Hof research station (48.74°N, 8.93°E; 478 m a.s.l.), in Southwest Germany on a loamy clay (28 % clay, 12 % sand and 60 % silt) in 2018. The 30-year average annual temperature and precipitation are 8.4 $^\circ$C and 738 mm.   The treatments (Table 2.8) were carried out with three replications (10 × 3 m^2) in a randomized complete block design, with 93,200 maize seeds ha^{-1}, seeded in 7 cm depth, using a pneumatic precision planter (Maxima 2TI, Kuhn Maschinen-Vertrieb GmbH, Schopsdorf, Germany).   The row distance was 75 cm and the intra-row plant distance was 16 cm. The herbicide and safener applications were carried with a field plot sprayer (Schachtner GmbH, Ludwigsburg, Germany) using flat fan nozzles

2.2. *2nd Study: Sensor-based Evaluation of Maize (Zea mays) and Weed Response to Postemergence Herbicide Applications of Isoxaflutole and Cyprosulfamide Applied as Crop Seed Treatment or Herbicide Mixing Partner*

(IDK 120-02, Lechler GmbH, Metzingen, Germany) simulating a carrier volume of 200 L ha^{-1}.

Data Collection

Visual Plant Rating and Biomass Collection: The plant damage was assessed visually at 8 days after treatment (DAT) as proposed by Robinson Robinson *et al.* (2013). The above-ground fresh matter of all greenhouse and field trial plants was collected 21 DAT and dried for three days at 80 $°$C. In the field trial, the above-ground biomass weight of the final crop harvest was recorded at 118 days after sowing. The species-specific weed coverage was estimated immediately before the herbicide application and 14 DAT with four repetitions per plot using a 0.1 m^2 counting frame.

Spectral Data Collection: The crop responses to herbicide and safener treatments were assessed by measuring the plant reflection changes with a HandySpec Field$^®$ (tec5 AG, Oberursel, Germany) portable hyperspectral sensor and multispectral cameras. The hyperspectral sensors spectra ranged from 360 to 1000 nm with a spectral resolution of 10 nm. Each point measurement was focused on the youngest unfolded maize leaf at a distance of 10 cm. Each treatment was measured 1, 2, 4, 6, 8 and 12 DAT at 14:00 UTC, $\pm$ 1 h, with five measurement repetitions per plant. Prior to the measurements, the spectrometer was calibrated with a BaSO$_4$ white reference.

Multispectral images were taken 8 DAT of each treatment with a Tetracam μMCA 4 Snap (Tetracam Inc., Chatsworth, CA, USA) multispectral camera in the greenhouse trial and a UAV mounted multispectral camera in the field trial. The camera gathered four narrow-band images at the spectral wavelengths of 550, 670, 780 and 900 nm using optical band-pass filters (10,10,10 and 20 nm bandwidth respectively), mounted in front of each of the camera's four image sensors. Images were gathered with a camera to plant distance of 75 cm at a resolution of 1280 × 1024 pixels. Spectrometer and multispectral camera measurements were performed outside the greenhouse to measure under natural light conditions.

In the field trial, a Sequoia (Parrot, Paris, France) multispectral camera was mounted as an in-build ungimballed sensor on a multirotor UAV, model Bluegrass (Parrot, Paris, France), gathering images at four narrow-band wavelengths of 550, 660, 735 and 790 nm (40, 40, 10, and 40 nm bandwidth respectively). The UAV images were collected from a flight altitude of 50 m above-ground level, resulting in a ground sampling distance (GSD) of 4.71 cm.

Data Processing and statistics

The statistical data evaluation was performed using the statistic program R Version 3.3.1. (?) Model 2.10 was used to analyze the variance of influences of test factors as a function of application date and the UAV multispectral imagery. The visually estimated plant damage, the spectral wavelength composition of the maize plants and the biomass data were evaluated with Model 2.11 by calculating the effect for the single maize plants. Subsequently, a multiple mean value comparison was carried out using a Tukey HSD test with p = 0.05.

$$y_{ik} = \mu + \alpha_i + \gamma_k + \epsilon_{ik} \tag{2.10}$$

$$(i, k = 1, ..., t)$$

$$y_{ik} = \mu + \alpha_i + \beta_j + \gamma_k + \epsilon_{ijk} \tag{2.11}$$

$$(i, j, k = 1, ..., t)$$

where μ = mean, α = effect of ith treatment, β = influence of the jth maize plant, γ = effect of kth repetition and ϵ = residual error.

The weed control efficacy (WCE) of the different treatments in the field trial was calculated as proposed by Rassmussen (1991) (Equation 2.12)as:

$$Weed\ Control\ Efficacy\ (\%) = 100 - \frac{ds_t}{0.01 \times du} \tag{2.12}$$

where *ds* is the estimated weed density 14 DAT after the herbicide application of each treatment (*t*) and *du* is the estimated weed density of the untreated control plots.

Spectral data analysis: The hyperspectral sensor and camera data was analyzed through the calculation of VIs. The four multispectral wavelength-specific images per triggering were combined to a single precisely overlaying multichannel TIFF image using the software PixelWrench2 (Version 1.2.3.1, Tetracam Inc., Chatsworth, CA, USA). Prior to the statistical analysis, the multispectral greenhouse plant images were manually clipped to the plant pixels. The multispectral images of the field trial were calculated to a georeferenced orthomosaic image with the help of a structure from motion software (PhotoScan Professional Edition, Version 1.4.5, Agisoft LLC, St. Petersburg, Russia). Prior to image analysis, the crop rows were clipped by their pre-known sowing position to avoid interference with weeds outside the crop rows. The multispectral camera and the hyperspectral sensor data of the greenhouse trial were compared by a correlation of the CropSpec vegetation index (Equation 2.13)(Reusch *et al.*, 2010) and the Green Normalized Difference Vegetation Index (GNDVI, Equation 2.14) (Gitelson *et al.*, 1996). Due to the different channel composition of the multispectral camera used at the field trial, the evaluation of the sensor information was additionally carried out on the Normalized Difference Red Edge Index (NDRE, Equation 2.15) (Eitel *et al.*, 2011). The indices were calculated as follows. Where R denotes the leaf reflectance at the specific wavelength in nm given in the sub-index.

$$CropSpec = \left(\frac{R_{900}}{R_{780}} - 1 \right) \times 100 \tag{2.13}$$

$$Green\ Normalized\ Difference\ Vegetation\ Index^* = \frac{R_{790} - R_{550}}{R_{790} + R_{550}} \tag{2.14}$$

*The Parrot Sequoia multispectral camera used in the field trial used a red channel at 780 nm instead of the 790 nm given in the GNDVI formula.

$$Normalized\ Difference\ Red\ Edge\ Index = \frac{R_{790} - R_{735}}{R_{790} + R_{735}} \tag{2.15}$$

2.2.3 Results

The pre-emergence IFT and CSA greenhouse trials did not show significant differences in any assessments between the CSA seed treatments. Both germination tests did not show a variation in the germination rates including the untreated control, although the mean germination time of the seeds at 10 $^\circ$C was 48 h later. At 20 $^\circ$C there were slightly longer roots remarkable at the CSA treated seeds compared to the untreated control at 4 days after trial start. Regarding the plant height and biomass production, there were no significant variations between CSA seed treatment dosages, as well as between each CSA dosage and the IFT treatments. Only the untreated CSA-free control suffered from every IFT dose rate, expressed as significantly reduced plant height and biomass.

The germination test prior to the presented trials showed no significant impact of the seed treatments and storage on the germination rate.

Greenhouse ground-truth

The IFT applied treatments showed the typically bleaching symptoms of HPPD-inhibition to a different extent, at the assessment 7 DAT (Table 2.9). The biomass of the untreated control at the 2-, 4- and 6-leaf stage reached 2.5, 3.4 and 5.6 g pot^{-1} respectively. The visual and sensor-based results were in line and indicated the lowest plant stress for the treatments at the 4-leaf stage inside each the COMB and SPLIT group. The dry biomass differences did not follow these results and continuously decreased in the course of the experiment. The used measurement methods confirmed bigger plant stress or reduced biomass production at all development stages for the SPLIT treatments. Likewise, all treatments suffered from the IFT application and indicated lower vitality parameters in all test methods. Despite the higher biomass reduction of the COMB-4 and SPLIT-4 treatments compared to the COMB-6 and SPLIT-6 treatments, the visually assessed maize plant damage was lower and not even significantly different from the untreated control. Thus, the visual plant damage assessment 7 DAT did not always follow the changes in the above-ground biomass at 21 DAT.

TABLE 2.9: The visual plant damage 7 days after treatment and the above-ground dry biomass 21 days after treatment of each treatment of the greenhouse pot trial are given relative to the untreated control. Values in bold represent a significant difference based on Tukey's HSD test ($\alpha = 0.05$) between the treatment and the respective control of the same day after treatment.

Treatment[†]	Maize damage (%)	Dry biomass difference (%)
COMB-2	6	**- 60**
SPLIT-2	**20**	**- 81**
COMB-4	0	**- 27**
SPLIT-4	2	**- 50**
COMB-6	2	**- 22**
SPLIT-6	7	**- 42**
Untreated control[‡]	0	0

† Untreated control = no *isoxaflutole* or *cyprosulfamide* application; COMB = *isoxaflutole* + *cyprosulfamide* combined spray application; SPLIT = *isoxaflutole* spray application and *cyprosulfamide* seed treatment.

‡ At the plant leaf stages of 2, 4 and 6

Greenhouse spectrometry

The hyperspectral sensor clearly showed always higher reflectance values at 550 nm for all COMB and SPLIT treatments, while different results were observed at 780 nm (Figure 2.4). The multispectral camera confirmed these measurements except for the COMB-6 and SPLIT-6 treatments, even though it resulted in much smaller differences in the reflection values between the treatments. Comparing the hyperspectral sensor NDRE and CropSpec vegetation index results (Table 2.10) 8 DAT there were statistically significant differences between all treatments at the 2- and 4-leaf plant development stage applications with both indices. Both VIs confirmed the differences found in the visual plant damage assessments within the respective application date.

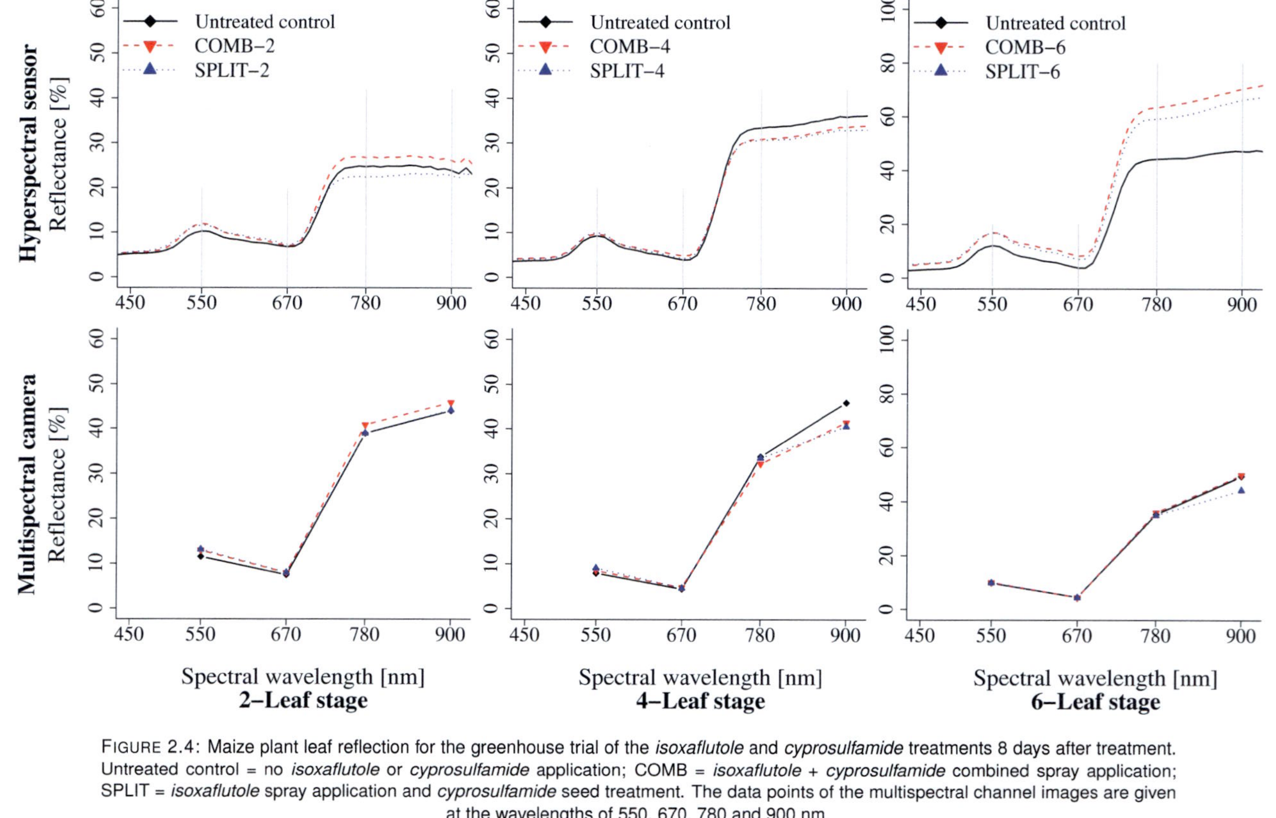

FIGURE 2.4: Maize plant leaf reflection for the greenhouse trial of the *isoxaflutole* and *cyprosulfamide* treatments 8 days after treatment. Untreated control = no *isoxaflutole* or *cyprosulfamide* application; COMB = *isoxaflutole* + *cyprosulfamide* combined spray application; SPLIT = *isoxaflutole* spray application and *cyprosulfamide* seed treatment. The data points of the multispectral channel images are given at the wavelengths of 550, 670, 780 and 900 nm.

TABLE 2.10: Variation of normalised difference red edge (NDRE) and CropSpec vegetation indices of maize plants at the greenhouse trial 1 to 12 days after treatment (DAT) at the 2, 4 and 6 leaf stage. Significant differences were calculated per plant leaf stage, day after treatment and vegetation index. Values with same letters indicate no statistical difference according to Tukey's HSD test (α = 0.05).

Treatment[†]	NDRE Vegetation Index						CropSpec Vegetation Index					
	1 DAT	2 DAT	4 DAT	6 DAT	8 DAT	12 DAT	1 DAT	2 DAT	4 DAT	6 DAT	8 DAT	12 DAT
Untreated Control-2	0.16^b	0.13^b	0.19^a	0.18^a	0.17^a	0.23^a	188^b	172^b	248^a	209^a	182^a	288^a
COMB-2	0.17^a	0.16^a	0.19^a	0.15^b	0.15^b	0.21^b	209^a	197^a	230^b	178^b	139^b	232^b
SPLIT-2	0.16^b	0.11^c	0.16^b	0.14^b	0.11^c	0.16^c	194^b	160^b	197^c	163^c	104^c	171^c
Untreated Control-4	0.17^a	0.15^a	0.23^a	0.26^a	0.27^a	0.28^a	180^a	183^a	283^a	346^a	330^a	340^a
COMB-4	0.18^a	0.14^{ab}	0.22^a	0.23^b	0.24^b	0.26^b	183^a	162^b	272^a	288^b	289^b	300^b
SPLIT-4	0.16^b	0.13^b	0.20^b	0.16^c	0.22^c	0.25^b	162^b	157^b	249^b	260^c	261^c	295^b
Untreated Control-6	0.27^a	0.27^b	0.28^a	0.27^a	0.23^a	-	337^a	350^a	337^a	326^a	278^a	-
COMB-6	0.27^a	0.28^a	0.28^a	0.26^b	0.22^b	-	335^a	331^a	335^a	316^b	245^b	-
SPLIT-6	0.25^b	0.25^c	0.26^b	0.25^c	0.23^a	-	327^a	297^b	312^b	288^c	261^a	-

† Untreated control = no *isoxaflutole* or *cyprosulfamide* application; COMB = *isoxaflutole* + *cyprosulfamide* combined spray application; SPLIT = *isoxaflutole* spray application and *cyprosulfamide* seed treatment.

The CropSpec calculated from the multispectral camera of the greenhouse trial correlated with an R^2 = 0.84 with the greenhouse hyperspectral sensor results. When calculated from the spectrometer, the CropSpec resulted in a lower index value compared to the camera data set, which was indicated by a correlation slope of + 0.757. The GNDVI showed a correlation of R^2 = 0.64 between the sensors and the slope was + 1.376 since the hyperspectral sensor-based GNDVI values were higher than the value of the same measurement calculated from the multispectral camera images.

Field ground-truth

In the vegetation period of 2018, the annual temperature and precipitation at the field location were 10.1 °C and 526 mm, with very dry weather conditions at the beginning of the trial and a heavy rain event with 35 mm m^{-2} 10 DAT after the first herbicide application. The untreated control of the field trial yielded a dry biomass 21 DAT of 0.5, 1.3 and 2.1 t ha^{-1} at the 2-, 4- and 6-leaf stage respectively (Table 2.11). As in the greenhouse trial, the SPLIT treatments showed increased visual plant damage and lower biomass production compared to the COMB treatments, with the exception of the SPLIT-2 treatment.

2.2. *2nd Study: Sensor-based Evaluation of Maize (Zea mays) and Weed Response to Postemergence Herbicide Applications of Isoxaflutole and Cyprosulfamide Applied as Crop Seed Treatment or Herbicide Mixing Partner*

TABLE 2.11: The visual estimated plant damage 7 days after treatment, the aboveground dry biomass 21 days after treatment and the herbicide efficacy 14 days after treatment of each treatment at the field trial are given relative to the control. The final dry biomass yield is given in absolute values. The weed control efficacy was assessed 14 days after treatment. Values in bold represent a significant difference based on Tukey's HSD test (α = 0.05) between the treatment and the respective control at the same days after treatment.

Treatment[†]	Maize damage (%)	Dry biomass difference (%)	Weed control efficacy(%)	Dry biomass yield (t ha^{-1})
COMB-2	**5**	- 16	**87**	13.52
SPLIT-2	**7**	- 10	**95**	13.80
COMB-4	**11**	+ 17	**92**	15.84
SPLIT-4	**23**	**- 25**	**96**	14.02
COMB-6	**18**	+ 29	**75**	14.67
SPLIT-6	**39**	**- 40**	**88**	7.91
Herbicide control	**48**	**- 72**	**96**	13.68
Untreated control[‡]	0	0	0	9.21

† Untreated control = no *isoxaflutole* or *cyprosulfamide* application; COMB = *isoxaflutole* + *cyprosulfamide* combined spray application; SPLIT = *isoxaflutole* spray application and *cyprosulfamide* seed treatment; Herbicide control = sole IFT spray application.

‡ At the plant leaf stages of 2, 4 and 6

Besides the weeds listed in Table 2.12 other, less frequently occurring, but important weeds were also found. These included: Catchweed bedstraw (*Gallium aparine* L.) and Black-bindweed (*Fallopia convolvulus* (L.) Á. Löve).

TABLE 2.12: Weed species composition of the field trial 14 days after treatment of the 3 three consecutive crop plant development stages (2, 4 and 6-leaf stage) of the untreated control given in % soil coverage. Other weed species with low coverage rates, were summarized and and given in the table as "others".

Botanical name	Weed species soil coverage (%)		
	2-leaf stage	4-leaf stage	6-leaf stage
Cirsium arvense	15	17	22
Galium aparine	0	2	3
Fallopia convulvulus	0	2	3
Chenopodium album	1	5	8
Polygonum aviculare	0	3	3
Polygonum lapathifolium	0	1	1
Matricaria perforata	0	4	4
Others	1	4	5
Total	17	38	49

UAV field spectrometry

The data of the hyperspectral sensor and the multispectral camera of the field trial showed the biggest differences between the treatments at 550 nm and above 780 nm, but also at 660 nm. Especially the herbicide control with the sole IFT application at the 4-leaf plant development stage resulted in a more flat reflectance curve in the hyperspectral sensor data and with values very close to the SPLIT-4 treatment in the multispectral camera dataset. The hyperspectral sensor-based NDRE and CropSpec indicated significant lower index values for the SPLIT treatments at the 4- and 6-leaf application dates (Table 2.13). For the COMB-2 treatment, both indices showed significantly reduced values over the entire measuring period. Already 1 DAT, there was no significant difference in the NDRE and CropSpec between the SPLIT-2 treated plants and the untreated control. The SPLIT-2 treatment resulted in significantly higher VI results and meanwhile showed less biomass reduction 21 DAT and at the final crop harvest compared to the COMB-2 treatment. These results were possible despite an 88% safener reduction compared to the COMB variants. As in the greenhouse trial, the hyperspectral sensor-based index calculation was in line with the visual estimated plant damage within the specific application dates.

2.2. 2nd Study: Sensor-based Evaluation of Maize (Zea mays) and Weed Response to Postemergence Herbicide Applications of Isoxaflutole and Cyprosulfamide Applied as Crop Seed Treatment or Herbicide Mixing Partner

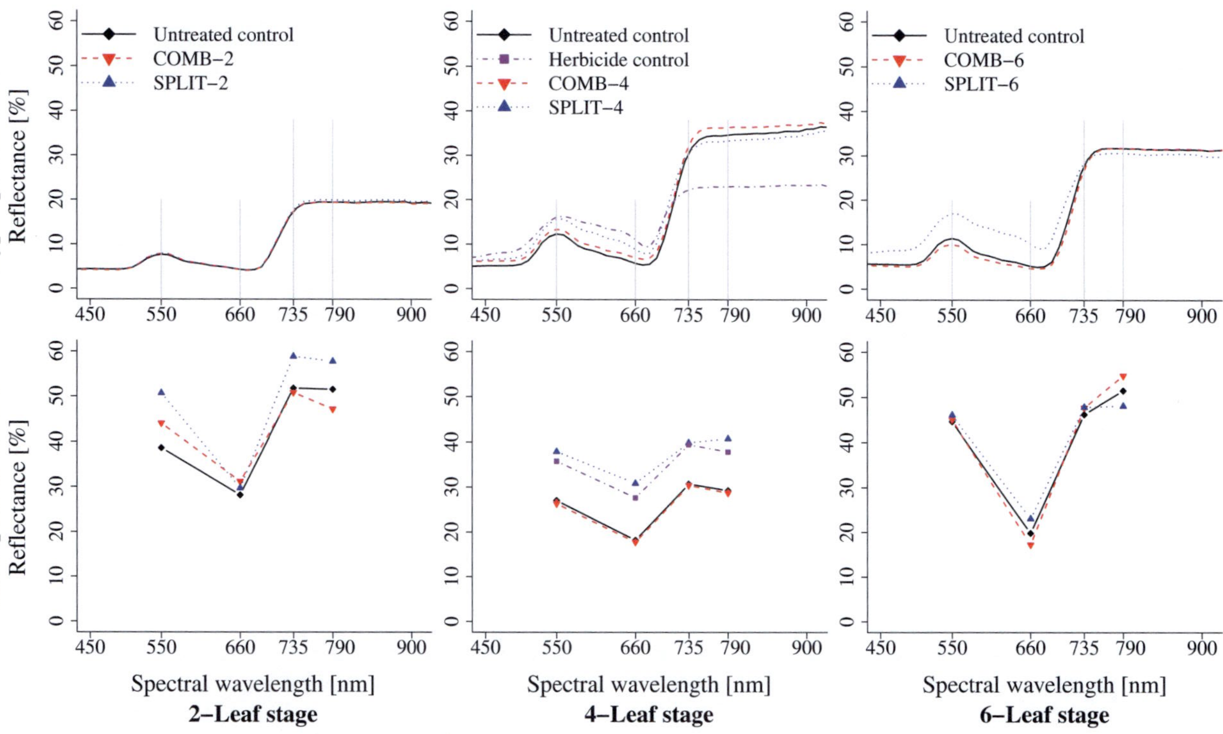

FIGURE 2.5: Maize plant leaf reflection for the field trial of the *isoxaflutole* and *cyprosulfamide* treatments at 8 days after treatment. Untreated control = no *isoxaflutole* or *cyprosulfamide* application; COMB = *isoxaflutole* + *cyprosulfamide* combined spray application; SPLIT = *isoxaflutole* spray application and *cyprosulfamide* seed treatment; Herbicide control = sole IFT spray application. The data points of the multispectral channel images are given at the wavelengths of 550, 670, 790 and 900 nm.

TABLE 2.13: Variation of normalised difference red edge (NDRE) and CropSpec vegetation indices of maize plants in the field trial 1 to 12 days after treatment (DAT) at the 2, 4 and 6 leaf stage. Significant differences were calculated per plant leaf stage, day after treatment and vegetation index. Values with same letters indicate no statistical difference according to Tukey's HSD test ($\alpha = 0.05$).

Treatment[†]	NDRE Vegetation Index						CropSpec Vegetation Index					
	1 DAT	2 DAT	4 DAT	6 DAT	8 DAT	12 DAT	1 DAT	2 DAT	4 DAT	6 DAT	8 DAT	12 DAT
Untreated Control-2	0.17^a	0.17^a	0.13^a	0.14^a	0.16^b	0.18^b	219^a	210^a	123^a	111^a	162^a	150^b
COMB-2	0.15^b	0.16^b	0.12^b	0.12^b	0.16^b	0.20^a	189^b	200^b	123^a	95^b	150^b	172^a
SPLIT-2	0.16^a	0.17^a	0.12^b	0.12^b	0.17^a	0.21^a	212^a	212^a	120^a	108^{ab}	162^a	180^a
Untreated Control-4	0.16^{ab}	0.16^a	0.15^a	0.16^a	0.20^a	0.19^b	163^a	159^a	138^a	148^a	212^a	168^b
COMB-4	0.14^c	0.15^b	0.13^b	0.15^{ab}	0.20^a	0.21^a	143^b	141^b	108^b	141^b	205^a	210^a
SPLIT-4	0.17^a	0.14^{bc}	0.12^c	0.14^b	0.13^b	0.15^c	161^a	123^c	97^b	138^b	138^b	129^c
Herbicide control	0.15^{bc}	0.13^c	0.12^c	0.13^c	0.05^c	0.04^d	114^b	128^{bc}	97^b	114^c	51^c	15^d
Untreated Control-6	0.16^a	0.19^{ab}	0.18^a	0.18^a	0.26^a	0.23^a	161^a	208^{ab}	161^a	167^b	280^a	275^a
COMB-6	0.10^b	0.20^a	0.16^b	0.19^a	0.26^a	0.23^a	59^b	216^a	130^b	192^a	276^a	271^a
SPLIT-6	0.11^b	0.18^b	0.14^c	0.15^b	0.13^b	0.15^b	75^b	201^b	113^b	128^b	124^b	167^b

† Untreated control = no *isoxaflutole* or *cyprosulfamide* application; COMB = *isoxaflutole* + *cyprosulfamide* combined spray application; SPLIT = *isoxaflutole* spray application and *cyprosulfamide* seed treatment; Herbicide control = sole IFT spray application.

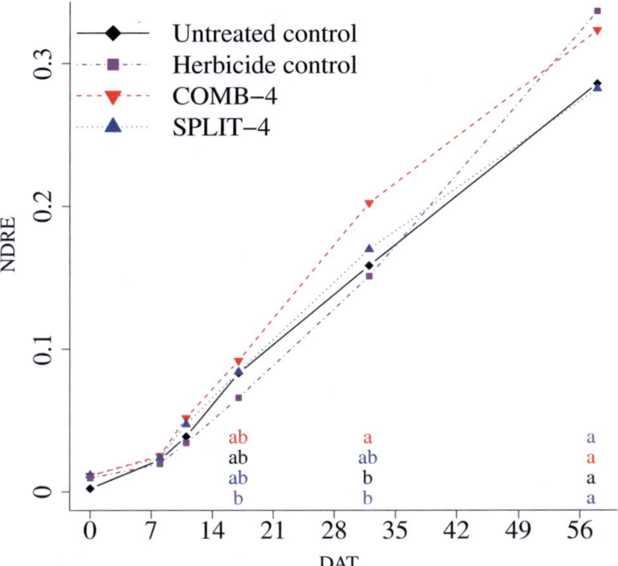

FIGURE 2.6: NDRE timeline calculated from the images collected with the multispectral camera mounted on the UAV 0, 8, 11, 17, 32 and 58 days after treatment of the COMB-4, SPLIT-4, untreated control and herbicide control in the field trial.Untreated control = no *isoxaflutole* or *cyprosulfamide* application; COMB = *isoxaflutole* + *cyprosulfamide* combined spray application; SPLIT = *isoxaflutole* spray application and *cyprosulfamide* seed treatment; Herbicide control = sole IFT spray application. Significant differences were observed only 17 and 32 days after treatment. The letters indicate significant differences based on Tukey's HSD test ($\alpha = 0.05$).

Employing the NDRE based on the hyperspectral sensor information, a separation of the treatments was possible already 1 DAT. Therefore, the initial better performance of the SPLIT-2 and 4 treatments until 4 and 2 DAT, respectively, became evident. In the UAV multispectral imagery-based VIs the first significant differences were observed in the flight 17 DAT of the 4-leaf plant development stage and were again statistically comparable 58 DAT. (Figure 2.6).

2.2.4 Discussion

In accordance with previous studies (Robinson *et al.*, 2013; Philbrook and Santel, 2008; Ahrens *et al.*, 2013), the greenhouse pot trial, as well as the field trial, confirmed the maize protection potential of the safener CSA in IFT treated plants to a different extent depending on the safener application method and application timing. The CSA treatments satisfactorily protected the maize plants of the COMB treatments in both trials and at all application dates. The protection potential of the SPLIT treatments did also show increased crop vitality directly after the IFT application. In particular, the plants of SPLIT treatments showed immediately after treatment, until 1 to 4 DAT less stress response to the IFT treatments. After this initial protection phase, the safener protection was apparently no longer sufficient. The safener effect also decreased later with treatment timings.

In the course of the greenhouse trial, the auditor estimated and sensor-based assessment of the deterioration of the maize plants decreased in the COMB and SPLIT treatments the later the date of treatment was set. Since IFT initially only bleaches leaves developing after treatment (Pallett *et al.*, 1998; Mayonado *et al.*, 1989), the ratio of damaged to healthy leaf parts decreases with progressive plant development. This leads to smaller biomass differences of the COMB and SPLIT treatments compared to the untreated control at the 4- and, even stronger, at the 6-leaf treatment dates. Therefore, the comparison of biomass production with auditor estimated plant damage and sensor-based CropSpec, GNDVI and NDRE shows that the two latter provide a more detailed indication of the plant health state and can even pinpoint differences (Table 2.9 and 2.11). However, the visually estimated crop injury did not necessarily indicate the safeners safening potential. Previous experiments found the IFT based crop injury to be transient and with no negative effect on maize height and total yield (Robinson *et al.*, 2013).

The slightly better WCE of IFT at the 4-leaf plant development stage application date may have been caused by the heavy rainfall event with 35 mm m^{-2} 3 DAT under the otherwise rather dry weather conditions during the field trial due to higher mobility and plant uptake of IFT (Taylor-Lovell *et al.*, 2000). IFT is relatively stable under dry soil conditions and to light and may tolerate heat for 14 days at 54 °C (Taylor-Lovell *et al.*, 2000; Tomlin, 2002). Thus, IFT can remain in the upper soil layers until a subsequent rainfall event causes a rapid transformation into the bioactive *diketonitrile* derivative and thus allows uptake by emerging plants (Taylor-Lovell *et al.*, 2000).

The inhibition of the *carotenoid* and thus bleaching of the leaves with a reduction of chlorophyll content in the meristematic tissues (Mayonado *et al.*, 1989) is directly related to plant vitality (Carter, 1998; Peñuelas and Filella, 1998) and can be assessed by hyper- or multispectral plant information (Peteinatos *et al.*, 2016). In this study, indices were chosen which are particularly sensitive to changes in the chlorophyll content of plants. The NDRE (Maccioni *et al.*, 2001) was already used to identify plant stress and thereby resulted in much higher accuracy (R^2 = 0.78) (Eitel *et al.*, 2010) than commonly used Normalised Difference vegetation Index (NDVI) (Rouse *et al.*, 1973). The CropSpec (Reusch *et al.*, 2010; Rasooli Sharabiani *et al.*, 2013; Reckleben, 2014) chosen in the greenhouse trial, was used to detect plant stress in winter wheat (Zecha *et al.*, 2018). The third VI employed in this study was the NDRE as a modification of the NDVI (Rouse *et al.*, 1973) and uses the green instead of the NDVI red term. The GNDVI was a suitable VI for quantifying plant health status since the green channel was used to improve chlorophyll related measurement (Schepers *et al.*, 2003). In contrast to the human assessment, sensors can provide more reliable and quantitative information of plant health, especially, when time-shifted treatments have to be compared (Andújar *et al.*, 2010). The NDRE and CropSpec VI results in this study showed significant and detailed information concerning different treatments within a measurement campaign and revealed differences in plant stress already 1 and 2 DAT, respectively. Hence, the VIs confirmed the visual estimated plant damage in both trials, except for the SPLIT-2 treatment.

The maize plant development in the field trial was slowed down by the herbicide application but showed succeeding recovery under favorable conditions. The further development of plants treated at the 4-leaf development stage, indicates recovery from the herbicide stress already 24 DAT, depending on the treatment (Figure 2.6). Another reason for the relative recovery of the higher stressed treatments can be a developmental depression in the less IFT-damaged plots. In these plots, corn plants are probably already competing for nutrients, water and light (Hasanuzzaman *et al.*, 2018). Due to the limited availability of these growth limiting factors in the IFT and CSA applications, plants could not capitalize on their improved herbicide tolerance. This may account for the immense growth of the CSA-free herbicide control in the field trial, which improved in final yield (Table 2.11) and UAV NDRE (Figure 2.6).

The correlation between the multispectral imagery and the hyperspectral sensor data illustrated the comparability of the results. Nevertheless, differences in the quality of the information can be seen especially in the multispectral aerial imagery. While the spectrometer was placed directly above the maize leaf to measure only the reflection of the maize, the multispectral camera collected the data from 50 m above the crop stand to be assessed in the field trial. The inevitable recording of ground and background pixels in the aerial images represented an important difference to the hyperspectral sensor data. This issue requires subsequent processing to separate plant and background/non-plant pixels prior to the evaluation. Even if established processing protocols could be used for this purpose, changes in the raw data must be carried out very carefully in order not to influence the result of the target variable, especially for the assessment of crop responses in hyperspectral datasets (Roth and Streit, 2017; Mink *et al.*, 2018). However, depending on the resolution and the crop development stage, in this study between the 8 and 10 leaf stage of the maize, the aerial multispectral imagery provided valuable crop stand information. From this stage onwards manual measurements, as with the hyperspectral sensor, would not have been possible any longer without causing plant damage by walking and working inside the crop stand.

2.2.5 Conclusion

CSA seed treatments reduced the amount of safener by 88 % compared to the traditional application method. This could be proved for all treatments at the field trial, without significant yield reductions. Therefore, the preferable crop stand development stage for the SPLIT application regarding crop stress response was at the 2-leaf stage in the field trial and the 6-leaf-stage in the greenhouse pot trial. The other SPLIT treatments showed higer plant stress compared to the COMB treatments and only recovered towards the end of the field trial. The applied spectral indices were suitable for plant health state assessments using hyperspectral sensors as well as UAV multispectral cameras. The general comparability of the sensor-based evaluations could be demonstrated in a correlation with an R^2 of 0.84 (CropSpec) and 0.64 (GNDVI).

Acknowledgements

The authors gratefully thank Markus Sökefeld for his assistance in the general trial planning and implementation. We would like to acknowledge the valuable contributions of the students Anna Dettweiler, Jan-Dilara Aydin and Lisa Paulus. The authors thank Deutsche Telekom AG, Bonn, Germany for providing the UAV including the mounted multispectral camera used at the field trial.

Conflicts of Interest

The authors declare no conflict of interest.

2.3 3rd Study: In-field Classification of Herbicide-resistant *Papaver rhoeas* and *Stellaria media* using an Imaging Sensor of the Maximum Quantum Efficiency of Photosystem II

Article

Published in Weed Research 59, 357– 366

Received: 23 July 2018 / Accepted: 05 June 2019 / Published: 15 July 2019

https://doi.org/10.1111/wre.12374

In-field classification of herbicide-resistant *Papaver rhoeas* and *Stellaria media* using an imaging sensor of the maximum quantum efficiency of photosystem II

by A. I. Linn, R. Mink, G. G. Peteinatos und R. Gerhards

1 Department of Weed Science, Institute of Phytomedicine, University of Hohenheim, 70599 Stuttgart, Germany

Summary

Reliable in-season and in-field tools for rapidly quantifying herbicide efficacy in dicotyledonous weeds are still missing. In this study, the maximum quantum efficiency of photosystem II (F_v/F_m) of susceptible and resistant *Papaver rhoeas* and *Stellaria media* populations in response to treatments with *textitacetolactate* synthase (ALS) inhibitors were examined. Seedlings (4 - 6 leafs) were transplanted into the field immediately after the application of the ALS-inhibitors *florasulam*, *metsulfuron-methyl* and *tribenuron-methyl*. The F_v/F_m values were assessed 1 - 7, 9 and 14 days after treatment (DAT). Based on the F_v/F_m values of all fluorescing pixels in the images of herbicide-treated plants, discriminant maximum-likelihood classifiers were created. Based on this classifier, an independent set of images were classified into 'susceptible' or 'resistant' plants. The classifiers' accuracy, false-positive rate and false-negative rate were calculated and presented. The F_v/F_m values of sensitive *P. rhoeas* and *S. media* plants decreased within 3 DAT by 28 - 43 %. The F_v/F_m values of the resistant plants of both species were 20 % higher compared to the sensitive plants in all herbicide treatments. The classifier separated sensitive and resistant plants 3 DAT with accuracies of 62 - 100 %. False-positive and false-negative classifications decreased with increasing DAT. We conclude that by the assessment of the F_v/F_m value in combination with the classification sensitive and resistant *P. rhoeas* and *S. media* populations could be separated 3 DAT. This technique can help to select effective control methods and speed up the monitoring process of susceptible and resistant weeds.

Keywords: ALS-inhibitor; decision support system; discriminant analysis; pulse amplitude modulated imaging fluorometer; herbicide resistance management

2.4 4th Study: Discrimination between herbicide resistant and sensitive *Stellaria media* plants using hyperspectral plant reflectance and artificial neural network weight analysis

Article
Submitted to Weed Research; ISSN:1365-3180

Discrimination between herbicide-resistant and sensitive *Stellaria media* plants using hyperspectral plant reflectance and artificial neural network weight analysis

Running title: Artificial neural network herbicide resistance detection

by Robin Mink, Alexander Ingo Linn, Roland Gerhards and Gerassimos Georgiou Peteinatos

Department of Weed Science, Institute of Phytomedicine, University of Hohenheim, 70599 Stuttgart, Germany

Summary

The development of field test systems for fast and reliable identification of herbicide-resis weeds are of great importance for weed management systems. In this study, the hyperspectral reflectance spectra of *acetolactate*-synthase (ALS) inhibitor resistant and sensitive *Stellaria media* plants were determined with a passive hyperspectral sensor (350 - 1000 nm, resolution 10 nm) following the application of the ALS-inhibitors *florasulam, metsulfuron-methyl* and *tribenuron-methyl* in three field experiments. An artificial neural network (ANN) was created to classify single plants to the classes 'resistant' or 'sensitive' based on their spectral reflectance following herbicide treatment. Furthermore, the weights of neurons in the ANN were analysed to subdivide the 65 input spectra according to their importance for discriminating sensitive and resistant plants. Subsequently, the 16 wavelengths with highest weights were used for training and testing a second ANN (redANN). The ANN was trained with unfiltered raw data of all 65 input spectra of each measurement gathered at one trial site. The validation of the ANN was performed with independent datasets from two experiment repetitions of two years. Within 4 to 7 days after herbicide treatment, depending on the test dataset, the ANN classified 72 to 95 % of the plants correctly. The redANN revealed equivalent results as the ANN with full input spectra.

Keywords: ALS-inhibitor; artificial intelligence; decision support system; herbicide resistance management; hyperspectral sensor

2.4.1 Introduction

The development of reliable in-field test systems for the identification of herbicide-resis weeds is of great importance. More than 260 herbicide-resistant weed species have evolved worldwide to 23 of 26 known herbicide sites of action (Heap, 2019). One of these weeds is *Stellaria media* (L.) Vill. (Common chickweed) with reported resistance cases against *acetolactate*-synthase (ALS)-inhibitors, but also synthetic auxins and photosystem II inhibitors (Heap, 2019). *Stellaria media* is a strong competitor for water, nutrients and space in cropping systems and germinates throughout the year (Wilson *et al.*, 1995; Turkington *et al.*, 1980). Target-site ALS resistant *S. media* often show no response to ALS-inhibitors (Marshall *et al.*, 2010; Linn *et al.*, 2019). The further development of the herbicide resistances in weeds may even jeopardise current efforts to reduce herbicide use, as higher doses or additional weed control measures become necessary (Kudsk and Streibig, 2003). Therefore, a rapid in-field detection of herbicide efficacy is important to adjust weed control measures within the current season.

Artificial neural networks (ANNs) can nowadays be found in almost every digitally-enabled sphere of life to analyse, support and accelerate daily tasks, based on immense databases. In weed science, plant identification, weed mapping and crop/weed population modelling are important domains of ANNs (Torra *et al.*, 2018; Chantre *et al.*, 2018; Mansourian *et al.*, 2017; Kamilaris and Prenafeta-Boldú, 2018).

Depending on the weed species and resistance case, reliable and quantifiable prediction of herbicide-resistance in a target plant can be assessed by sensor-based in-field tests, as they provide auditor independent results prior to visual appearance (Wang *et al.*, 2016; Linn *et al.*, 2019). While in several studies chlorophyll fluorescence sensors were used for herbicide resistance detection in weeds (Ahrens *et al.*, 1981; Norsworthy *et al.*, 1999; Mechant *et al.*, 2010; Kaiser *et al.*, 2013; Linn *et al.*, 2018, 2019), fever studies used hyperspectral sensors for this purpose (Reddy *et al.*, 2014; Lee *et al.*, 2014; Nugent *et al.*, 2018).

Reddy *et al.* (2014) demonstrated the discrimination of *glyphosate*-resistant and *glyphosate*-susceptible *Amaranthus palmeri* S.Wats. (Palmer amaranth) by hyperspectral imagery for mixed laboratory and field plants. To identify herbicide sensitive and resistant *A. palmeri* plants prior to *trifloxysulfuron-methyl* application Matzrafi *et al.* (2017) evaluated hyperspectral reflectance profiles and achieved an accuracy of up to 90.9 %. Lee *et al.* (2014) could differentiate between *glyphosate*-susceptible and *glyphosate*-resistant *Lolium multiflorum* Lam. (Italian ryegrass) biotypes, also by hyperspectral data.

The still low implementation of hyperspectral sensors and hyperspectral data for the detection of herbicide resistance under field conditions contrasts with the otherwise versatile applications of spectroscopy in life sciences and especially in agriculture. Passive hyperspectral sensors can provide highly detailed information concerning crop information (Geipel and Korsaeth, 2017), crop vigour (Rumpf *et al.*, 2010; Mahlein *et al.*, 2012, 2018), and weed infestation (Yan Huang *et al.*, 2001; Thorp and Tian, 2004). Thereby the sensors are operated on different sensor carriers as satellites (Soria-Ruíz and Fernández-Ordoñez, 2003) UAVs (Zaman-Allah *et al.*, 2015) and ground-based sensors (Peteinatos *et al.*, 2016). A major reason for the comparably small impact of hyperspectral sensing for herbicide resistance detection might be due to the high dependency of these data on light quality. Therefore, the measurement day, the time dependency for the light conditions, as well as the plant dependent reflectance quality may affect the measurement results (Peteinatos *et al.*, 2016). These interferences often prevent the data from being used for statistical evaluation or may lead to considerable influence on the measurement results.

Here, ANNs can be used to recognise patterns in big datasets, which would be otherwise unnoticed and often offer a potential solution to elicit reliable conclusions from complex and tremendous amounts of data. Thus, ANNs are well known to indicate very subtle functional relationships even if these cannot be explained directly or correlated by normal models (Zhang *et al.*, 1998). Keränen *et al.* (2003) have used ANNs for automated plant identification using chlorophyll fluorescence fingerprinting. They also highlighted the suitability of ANNs for plant identification purposes with highly promiscuous data. The sensors used for the collection of hyperspectral plant vigour information are expensive, especially when it comes to multi- or hyperspectral cameras. Thus, pre-knowledge on the wavelengths relevant for specific study objectives can facilitate the choice of the sensor or sensor-setup. This knowledge can be enhanced through ANN weighting analysis. Gorman and Sejnowski (1988) performed a weight pattern analysis on a neural network trained by sonar data for object identification. Thereby the authors of that study identified specific signal frequencies, critical for the object identification. Chen *et al.* (2014) used ANNs to learn deep features of hyperspectral data and create auto encoders for these data. Thus, ANNs could not only be trained and used for classification applications but were already analysed for the identification of relevant data input. This method may provide important information if applied on weed response after herbicide treatment and thus may reveal essential inputs critical for herbicide resistance detection. Based on this information future herbicide resistance classification systems can be designed. Accordingly, we hypothesise that a) ANNs can discriminate between resistant and sensitive *S. media* plants based on their hyperspectral reflections if they have been trained in advance with a corresponding training dataset; b) the identification is also possible if the verification dataset originates from a completely independent experiment (plant material, year and trial site); c) the following analysis of the ANN neuron weighting can indicate wavelengths of high impact on the classification accuracy.

The objectives were i) the training of an ANN with hyperspectral spectrometer plant reflectance measurements of *florasulam, metsulfuron-methyl* and *tribenuron-methyl* treated ALS resistant and sensitive *S. media* plants in a field experiment; ii) the subsequent validation of this ANN using independent datasets collected at a second field site in two vegetation periods; iii) the identification of classification critical spectra by neuron weight analysis and the subsequent validation of their importance by the creation of ANNs only trained and tested with the identified 'important' and 'non-relevant' spectra.

2.4.2 Materials and Methods

Plant Material and Herbicide Treatment

Seeds of two populations of *S. media* with confirmed ALS-inhibitor resistance (STEME-R) and sensitivity (STEME-S), were obtained from INDENTXX (IDENTXX GmbH, Stuttgart, Germany) and Herbiseed (Herbiseed Ltd., Berkshire, UK) as used and tested in a concurrent study by (Linn *et al.*, 2019). The test of herbicide resistance in the STEME-R population revealed a single nucleotide polymorphism (SNP) on the ALS gene (Linn *et al.*, 2019). The plants were pre-grown in vermiculite (Perleen 437-2; Rench Chemie GmbH, Renchen, Germany) in the greenhouse at 20/15 $^\circ$C $\pm$ 4 $^\circ$C day/night temperature, with a relative humidity of 70 % and a 12 h photoperiod with 400 µmol m^{-2} s^{-1} light. When the plants had two leaves, these seedlings were transplanted into multipot plates (Anzuchtplatte Premium 84; Green24 GmbH, Bochum, Germany) filled with cultivation soil (Pikiererde CL P Classic; Einheitserde Werkverband e.V., Sinntal-Altengronau, Germany). At the 4- to 6-leaf stage, the herbicides were applied as given in Table 2.14.

TABLE 2.14: Herbicide treatments applied to *Stellaria media* (L.) Vill.

Herbicide active ingredient	Trade name	Active ingredient rate[†]	Formulation[‡]	Application rate	Provider
florasulam	Primus®	50 g L^{-1}	SC	0.1 L ha^{-1}	Dow Agroscience
metsulfuron-methyl	Allie® SX®	200 g kg^{-1}	SG	40 g ha^{-1}	Du Pont de Nemours
tribenuron-methyl	Pointer® SX®	500 g kg^{-1}	SG	40 g ha^{-1}	Du Pont de Nemours

† Spraying volume: 200 L water ha^{-1}.

‡ SC, suspension concentrate; SG, water-soluble granules.

The application was carried out in a precision laboratory track sprayer (Detron AG, Stein, Swiss) using a single flat fan nozzle (8002EVS; TeeJet®, Spraying Systems Co., Wheaton, IL, USA), at a simulated application volume of 200 L ha^{-1} and speed of 800 mm s^{-1}. The untreated control plants were treated the same way but with water only. The plants were transplanted into the field two hours after the herbicide application.

Field Trials

The field experiments were located in Southwest Germany, at the research stations Ihinger Hof (IHO, 48.74°N, 8,92°E; 478 m a.s.l.) in 2017 and at Heidfeldhof (HFH, 48.42°N, 9.11°E; 400 m a.s.l.) in 2017 and 2018. The average 30-year annual temperature and precipitation are 8.4 °C and 738 mm at IHO and 8.5 °C and 685 mm at HFH. There was no rain at both trial sites and both vegetation periods for 24 h, after herbicide treatment and transplanting of the plants into the field. The field trials were set up as a Latin square split-plot design. The herbicide treatments and the untreated control were attributed as the Latin square factor, while the populations (STEME-R and STEME-S) as the sub-factor. Three plants of each population were transplanted into each plot (1.5 × 3 m^2).

Spectrometer Measurements

From 4 to 7 days after treatment (DAT) plant reflection of each plant was measured with the HandySpec Field® (tec5 AG, Oberursel, Germany) hand-held passive spectrometer once per day at each respective trial site, with five measurement repetitions per plant. The spectrometers spectra ranged from 360 to 1000 nm with a spectral resolution of 10 nm, resulting in 65 measured spectra. This spectrometer consists of two independent sensors, with one pointing downwards to measure the leaf reflection and the second pointing upwards to capture the down-welling radiation through a cosine diffuser. Each point measurement was focused on the youngest leaf from a distance of 5 cm and was five times repeated. Prior to the measurements, the spectrometer was calibrated with a white standard ($BaSO_4$) as reference.

Artificial Neural Network Training and Classification

An ANN has been created and trained on the collected data. The ANN used as inputs all 65 spectral values provided from the spectrometer at the input layer. The target of the ANN was to classify between sensitive and resistant plants, after the herbicide treatment, therefore these two classes were implemented as the outputs of the network. Two hidden layers were added prior to the output. The first hidden layer was a dense layer of 25 nodes using a rectified linear unit (ReLU) for activation (Goodfellow *et al.*, 2016). The ReLU function is defined as following:

$$f(x) = \begin{cases} x, & \text{if } x > 0; \quad \text{with } x \in \mathbb{N} \\ 0, & \text{otherwise} \end{cases} \tag{2.16}$$

The second hidden layer was a dense layer of 10 nodes also using ReLU as its activation function. A 50 % node dropout at each update during the training was applied, in order to reduce the over-fitting and the generalisation error (Srivastava *et al.*, 2014).

For the training of the ANN the training data had been augmented as follows: for each unique combination of population and treatment, all the measurements at the same date were pooled together. Apart from the original measurements, a mean of each possible measurement combination in the pool was created and added in the training data, attributed to the same population and treatment. For the validation of the ANN, only the original measurements gathered have been used without any augmentation. The data from the location HFH in 2017 have been used for the training of the ANN, while the data from the same location in 2018 and the location IHO in the year 2017 have been used for the validation of the ANN.

The values predicted on the validation datasets were compared with the classes annotated to each plant reflectance measurement. The accuracy of the ANN was calculated as indicated in equation 2.17 where TP denotes the true positively identified plants, FP the false positively identified plants, TN the true negatively identified plants, and FN the false negatively identified plants (Witten *et al.*, 2011). Furthermore, for each of the two classes, the precision of the class was calculated as indicated in equation 2.18 and the recall as indicated in equation 2.19. Precision measures the correctly classified plants out of all the plants that were classified in the specific class, while recall gives the correct prediction value out for that specific plant class. Additionally, the false-negative (Equation 2.20) and the false-positive (Equation 2.21) rates were calculated.

$$Accuracy = \frac{TP + TN}{TP + FP + TN + FN} \tag{2.17}$$

$$Precision = \frac{TP}{TP + FP} \tag{2.18}$$

$$Sensitivity = \frac{TP}{TP + FN} \tag{2.19}$$

$$False - Negative\ Rate = \frac{FN}{TP + FN} \tag{2.20}$$

$$False - Positive\ Rate = \frac{FP}{FP + TN} \tag{2.21}$$

Artificial Neural Network Neuron Weight Analysis and Input Reduction

To identify the impact of each input spectra on the classification, a neuron weight analysis has been performed on the full spectra ANN, used for the classifications. To this end, the connections between the neurons of the first hidden layer and the input layer were analysed. During the training process (parameterisation) of the ANN, each neuron optimises the weight of each input received to maximise the ANN's accuracy. Therefore, each neuron of the first hidden layer has refined the weight of it's connection to each input spectra during the training. With increasing importance of a specific input spectra, the absolute numerical value of the connection of the associated neurons also increases. In order to determine the importance of a spectra (x) on the ANN performance, the square sums for the values (weights) of the individual connections between input layer and neurons (n) of the first hidden layer were determined.

$$sum\ of\ squared\ weights(x) = \sum_{i=1}^{n} (weight(x))^2 \tag{2.22}$$

The square sums of the weights (SSW) have been sorted beginning with the highest SSW and subsequently partitioned in three subsets by determining the quantiles. The first subset (Q1) consisted of the 16 input spectra of the 1st quantile with the highest SSW. The second subset (Q3) consisted of 16 input spectra of the 3rd quantile, with the median as the highest SSW within Q3. The third subset consisted of the remaining spectra.

To proof the relevance of the spectra, two additional ANNs (Q1 ANN, Q3 ANN) were created and trained for 100 thousand epochs using the identified spectra of Q1 and Q3, respectively. Due to the reduced number of input spectra, the Q1 ANN and Q3 ANN consisted only of 6 neurons in the first and 3 neurons in the second hidden layer. Apart from that, the architecture of the Q1 ANN and Q3 ANN was similar to the full spectra ANN. The validation was done also on the reduced Q1 and Q3 datasets, respectively.

2.4.3 Results

In the field trials of both growing seasons, all untreated control plants and herbicide-resistant plants survived the transplantation and herbicide treatments. The herbicide treated sensitive plants were finally dead 28 DAT. The untreated control, as well as the herbicide treated resistant plants, did not show visual stress symptoms after transplantation.

Training and Validation

Averaged over 4 to 7 DAT, the ANN classified the validation data of 2017 with an accuracy of 95 % and false-positive and false-negative rates below 5 % (Table 2.15). The classification accuracies varied between DAT and the herbicide treatment (Table 2.16). The highest accuracies were achieved 6 DAT for both florasulam and metsulfuron-methyl with 100 %. For tribenuron-methyl, the classification accuracies were generally lower with the highest value at 5 DAT with 96 %. In 75 % of the cases, the false-positive rate was below 5 %, while 58 % had a false-negative rate below 5 %.

The ANN's accuracy, precision and sensitivity for the test data of 2018 were lower compared to 2017. Accordingly, false-positive rate and false-negative rate increased. Here, the false-positive rate was three times higher than for 2017, while the false negative rate increased more than 14 times. Thus, only 16 % of true resistant plant measurements were attributed to the wrong class.

TABLE 2.15: Artificial neural network classification performance indicators for the detection of resistant *Stellaria media* plants pooled from 4 to 7 days after herbicide treatment in the vegetation periods of 2017 and 2018.

Vegetation period	Accuracy (%)	Precision (%)	Sensitivity (%)	False-negative rate (%)	False-positive rate (%)
2017	95	97	95	5	4
2018	72	78	84	16	58

The classification validation is based on the hyperspectral plant reflectance. Training was done with the datasets from the trial station Heidfeldhof 2017, while the validation was conducted with the datasets from Ihinger Hof 2017 and Heidfeldhof 2018.

TABLE 2.16: Herbicide and daily classification accuracy of the artificial neural network for the detection of resistant and susceptible *Stellaria media* plants 4 to 7 days after herbicide treatment in the vegetation period of 2017.

Herbicide	Accuracy (%) Day after treatment			
	4	5	6	7
florasulam	91*	98*‡	100*‡	96*‡
metsulfuron-methyl	95‡	99*‡	100*‡	96‡
tribenuron-methyl	94*	96*	92*	82

The classification is based on the hyperspectral plant reflectance. Training was done with the datasets from the trial station Heidfeldhof, while the validation conducted with the datasets from Ihinger Hof.

* False-positive rate ≤ 5 %
‡ False-negative rate ≤ 5 %

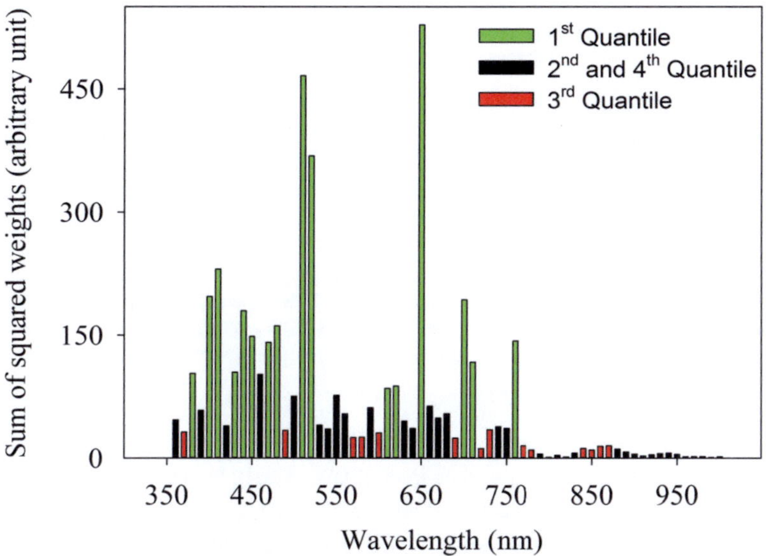

FIGURE 2.7: Sum of squared weights of the neuron weight analysis results per spectra. Green marked bars indicate spectra of the 1st quantile, spectra with red bars belong to the 3rd quantile. The spectra of the 2$_{nd}$ and 4$_{th}$ quantiles are given in black.

Neuron Weight Analysis

The neuron weight analysis allowed the identification of important, classification relevant input spectra by their overall weighting in the ANN (Figure 2.7). Ten of the 16 wavelengths with the highest SSW (Q1) were located between 380 and 520 nm. The wavelength with the highest SSW was at 650 nm. In addition, the reflectance spectra at 610 and 620 nm, 700 and 710 nm and 760 nm as well belong to the Q1 spectra. However, the most important spectra were found at 500, 510 and 640 nm and no high weights were shown above 760 nm and between 530 and 610 nm.

The validation of the Q1 ANN with test data of 2017 showed almost identical results to the full spectra ANN (Table 2.17). Averaged over 4 to 7 DAT, the Q1 ANN classified the test data with an identical accuracy as the full spectra ANN (Table 2.15). In 2018 the Q1 ANN classified 8 % more accurate with 78 % than the full spectra ANN on the same test data. Also the precision of the Q1 ANN increased by 1 % in 2018 and dropped from 97 % to 96 % in 2018 compared to the full spectra ANN.

TABLE 2.17: Artificial neural network classification performance indicators for the detection of resistant *Stellaria media* plants pooled from 4 to 7 days after herbicide treatment in the vegetation periods of 2017 and 2018 for selected wavelengths, based on the neuron weight analysis from a previous trained artificial neural network.

Artificial neural network	Accuracy (%)	Precision (%)	Sensitivity (%)	False-negative rate (%)	False-positive rate (%)
Q1 2017	95	96	95	5	6
Q1 2018	78	79	97	3	83
Q3 2017	76	71	100	0	60
Q3 2018	76	76	100	0	100

The classification is based on selected wavelengths of the hyperspectral plant reflectance. Training was done with the datasets from the trial station Heidfeldhof 2017, while the validation was conducted with the datasets from Ihinger Hof 2017 and Heidfeldhof 2018.

The Q3 ANN showed lower performance than the full spectra ANN and the Q1 ANN with all test data. With test data from 2017 and 2018 the Q3 ANN showed an accuracy of 76 %. For both test datasets, the Q3 ANN had a sensitivity of 100 %, and a false-positive rate of 0 %, since all measurements of true sensitive plants were assigned to the class 'resistant'.

2.4.4 Discussion

The ANN was used as a pre-trained classifier for the assignment of herbicide treated *S. media* plants to the classes 'resistant' and 'sensitive' by shifts in the plant reflection spectra. One of the major problems of hyperspectral plant reflectance data is the lack of robustness of the prediction in different environmental measurement conditions (Reddy *et al.*, 2014; Peteinatos *et al.*, 2016).

To this end, the ANN classifier provided a suitable solution to overcome classical statistical data comparison. Thus, the ANN tries to identify and assign the measurement to one of the pre-trained groups, which is, in this case, one of two possible output layers. Therefore, single measurements were evaluated, which prevents the misclassification of a population due to some outliers in the datasets to be compared. With accuracies of 95 % of the full spectra ANN in 2018 and even 78 % of the Q1 ANN in the completely separated vegetation period of 2017, the suitability of the experimental setup for in-field, in-season herbicide resistance detection using pre-trained ANN classifiers was demonstrated. Already the classification accuracies of 100 % at 6 DAT for the *metsulfuron-methyl* and *florasulam* treatments using a completely different dataset from different plants at a second location were remarkable. At least for the investigated ALS-inhibitor resistant population of *S. media* with the tested herbicides, only a minor dependency of the data on the vegetation period was found. Therefore, we expect that the ANN could be used to classify sensitive and resistant weed species also in other fields and seasons. It should be highlighted that the ANN of this study was specifically trained with unfiltered data. This ensures that the learning method and the consequent classification result was not based on and affected by pre-selected training datasets. It is possible that the accuracy of the ANNs could be further improved by a more detailed examination and filtering of the training data values and higher resolution or accurate sensor systems (Reddy *et al.*, 2014).

Thereby, the method was chosen in such a way that it can be easily applied and tested to other species and herbicides. Thus, the system may also be suitable for use by farmers and consultants as the measurement can be done rapidly and the resulting amounts of data are very small. These can be easily transmitted via internet and evaluated in cloud-based ANNs. This would allow additional weed control measures to be carried out directly in-season, to avoid further infestation of the field with herbicide-resistant plants.

The previous presented concurrent study by Linn *et al.* (2019) demonstrated already the classification potential with a supervised maximum-likelihood classifier by the use of the more distinct active chlorophyll fluorescence measurement. In this study, the chlorophyll measurements were thereby found to give in-field classification results with up to 100 % accuracy already at 3 DAT. Nevertheless, the measuring procedure with the active pulse amplitude modulating imaging fluorometer is comparatively complex and tedious. By transferring this principle to an ANN, hyperspectral measurements of plant reflection could significantly reduce the measurement time. Hyperspectral measurements have the advantage that they can be carried out without a prior dark adaption of the plants, and therefore imaging sensors could potentially be used to cover more plants on a larger area, compared to a "one-pixel" point spectrometer.

While the test system presented in this work was able to work under field conditions and at different locations and vegetation periods with classification accuracies between 72 % and up to 100 %, the following studies presented comparable results, but with several system limitations. The study conducted by Reddy *et al.* (2014) using hyperspectral plant imagery, achieved a classification accuracy of 94 % for greenhouse-grown, and 96 % for field-grown *A. palmeri* mixed genetically homogeneous and heterogeneous plants at the same vegetation period. Matzrafi *et al.* (2017) used a leaf clip hyperspectral sensor to measure *A. palmeri* pot plants under artificial light conditions and was able to distinguish between resistant and susceptible plants with an accuracy of 90.9 %, prior to application of *trifloxysulfuron-methyl*. The differentiation of *glyphosate* susceptible and resistant *L. multiflorum* plants was also possible with accuracies of up to 80 % using hyperspectral imagery (Lee *et al.*, 2014).

With a precision on the target 'resistant' of 97 % in 2017 and 78 % in 2018 the full spectra ANN showed similar results as the findings presented in Linn *et al.* (2019). The day and herbicide wise comparison as shown in table 2.16 showed slightly better results than the presented in the concurrent study (Linn *et al.*, 2019). The precision is especially important in order not to overlook any resistant plants and thus risk further spreading of herbicide-resistant seeds. Nevertheless, it should be added that the low false-negative values are not the only decisive factor for the suitability of a classifier, which is shown when considering the Q3 ANN. But the low false-positive rates in combination with a high accuracy, as seen in the Q1 ANN (Table 2.17) and full spectra ANN (Table 2.15 and 2.16) already at 4 and even better at 5 and 6 DAT, are suitable for the specific tracking of resistant individuals.

The analysis of neuron weights evolved through training of the full spectra ANN provided further information on the relevance of the measured wavelengths on the classification result. We identified a quarter of the spectra used to have higher neuron weights (Q1) compared to the other spectra (Figure 2.7). One-half of the Q1 wavelengths were found in the blue region from 370 to 470 nm, with the other half, spread over the green, red and near-infrared regions. The classification critical wavelengths identified by the neuron weight analysis were also found by Reddy *et al.* (2014) in similar regions. Reddy *et al.* (2014) show the critical wavelengths in the range of 394 – 487 nm, 668 – 762 nm and 821 – 863 nm. These authors also analysed the validation accuracy in relation to the number of input wavebands and found a plateau around 13 input wavebands, which is very close to the 16 input spectra used for the Q1 and Q3 ANNs. Using ANNs for improving hyperspectral data can be a promising use of this technology. Chen *et al.* (2014) have already tried to create autoencoders in order to improve hyperspectral data with promising results, yet this was just done to improve the quality of the provided data. Gorman and Sejnowski (1988) tried also to do an analysis of ANN hidden layers in order to pinpoint important wavelengths for a simple object separation.

In our case the reduction did not only reduce the number of input spectra but at the same time increased the accuracy of the Q1 ANN in 2018. This may be accompanied by a reduction in the dimensionality of the data as described in the Hughes phenomenon (Hughes, 1968). Since the number of statistical parameters increases with a growing number of variables (input spectra), the number of estimations on these statistical parameters also increases with increasing dimensionality of the data. Thus, as reviewed in Reddy *et al.* (2014), Hughes (1968) stated that the benefit of adding more input is outweighed at some point by the increased error in parameter estimates. The Q1 ANN, with only a quarter of the input, reached almost the same accuracy in 2017 and was even 6 % higher in 2018. This confirms the selection of the important spectra, critical for the separation of sensitive and resistant *S. media* plants.

For further confirmation of the correct selection of the Q1 spectra, the Q3 ANN was created. The high false-positive rates of the Q3 ANNs indicate the tendency of the presented classifiers, which were already found to a smaller amount in the Q1 ANN of 2018. The spectra of the 4th quantile were deliberately not used for the reduced ANNs since they were mostly in the near-infrared range, which are known to have a low signal to noise ratio due to the sensor and wavelength characteristics.

Through the comparison of the classification performance of the three created ANNs, the importance of specific relevant spectra becomes obvious. The identification of this spectra by the analysis of neuron weights aimed at a better understanding of the hyperspectral fingerprints and shifts of the examined plants. A further goal was to investigate the possibility of neural networks for the identification of plant stress or interactions in complex or multi-layered datasets, which were mostly investigated classically using vegetation indices. The approach presented was found to be suitable to detect the classification critical input spectra and may facilitate the selection in the search of an appropriate sensor for following studies.

2.4.5 Conclusion

The ANN using the reflectance spectra from 350 to 1000 nm in 65 spectra as data input, classified ALS resistant and sensitive *S.media* plants with an accuracy of 95 % and 72 % in the first and second vegetation period, respectively. The wavelengths critical for the classification of resistant plants could be identified by neuron weight analysis of the ANN. This improved the hyperspectral sensor for the specific task in herbicide resistance detection and lead to higher classification accuracies. The suitability of this parameter identification method could be proved through the creation of two following ANNs (Q1 and Q3 ANN) and comparison of the classification performance.

Funding

The project was supported by funds of the Federal Ministry of Food and Agriculture (BMEL) based on a decision of the Parliament of the Federal Republic of Germany via the Federal Office for Agriculture and Food (BLE) under the innovation support programme.

Acknowledgements

The authors gratefully thank Alexandra Heyn for her help in setting up and carrying out the study. We would like to acknowledge the excellent fieldwork of the students Rebekka Steinhart, Hannah Ertelt and Nico Wohlfarth.

Conflicts of Interest

The authors declare no conflict of interest.

Chapter 3

General Discussion

Precision farming (PF) applications target the site-specific optimisation of crop cultivation measures on the heterogeneity of biotic and abiotic parameters influencing crop development as well as the economic conditions (Heege, 2013). In weed management, PF aims to adopt control measures to the spatial and temporal variations of weeds (Marshall, 1988; Johnson *et al.*, 1995). Therefore the prerequisites of a PF approach can be structured into three steps: i) determination of the current field conditions (e.g. weed coverage) using sensors suitable for the specific task, ii) suitable evaluation to derive treatment parameters from the collected data, and (iii) farm machinery capable to conduct the site-specific treatment according to the calculated parameters of plants or sub-areas.

The implementation of a multitude of sensors in farming equipment has led to immense amounts of data, often increasing the monitoring and application complexity of crop cultivation measures (Stafford, 2000; Lindblom *et al.*, 2017). Nevertheless, the comprehensive field information already available from other PF applications can be analysed and combined with additional sensor data to use it for meaningful PF applications. Thereby UAVs equipped with hyperspectral and red-green-blue (RGB) cameras can provide information needed to monitor fields and enable site-specific weed management. In the studies presented, existing PF techniques were investigated and refined to extend the PF spectrum in weed management and to complement or modify existing PF weed management applications.

Site-specific weed management targets weed control according to the spatial and temporal heterogeneity of weeds (Gutjahr and Gerhards, 2010), to take into account local economic and ecological threshold levels. The use of UAV derived imagery for computing herbicide application maps has long been discussed in the literature (Rasmussen *et al.*, 2013; Peña *et al.*, 2013; de Castro *et al.*, 2018). The main aspect of site-specific herbicide applications is the reduction of herbicides to reduce costs, herbicide burden on the fields and the environment and thus the risk of herbicide residues in the food chain (Gerhards and Oebel, 2006; Gerhards, 2010; Jensen *et al.*, 2012). In the 1st study (Section 2.1) the "Multi-temporal Site-specific Weed Control of *Cirsium arvense* (L.) Scop. and *Rumex crispus* L. in Maize and Sugar Beet Using Unmanned Aerial Vehicle Based Mapping" is presented.

The known system limitations for UAV weed mapping, as well as alternative approaches are reviewed in Fernandez-Quintanilla *et al.* (2018). Therefore, the flight endurance of the UAVs, the sensor resolutions and the amount of data to be evaluated are the key limitations of UAV-based weed mapping for site-specific herbicide applications. In order to use UAV-based aerial imagery for site-specific herbicide applications, the computation of application maps must be adapted to cope with these limitations. Therefore, in this study, the amount of data to be calculated was reduced by a quick separation of vegetation and non-vegetation pixels (e.g. soil, mulch residues). For the computation of the weed maps, comprehensive PF data from other applications (e.g. sowing and digital field maps) was included. By removing the crop rows from the binarized vegetation maps, the amount of data could be reduced already at the beginning of weed maps creation. Compared to other image analysis methods such as object-based image analysis (OBIA), for this procedure, only low computing power was required. Thereby, a considerable loss of spatial resolution in the images was remarkable which can be a drawback for further investigations using weed pattern recognition.

In addition, the above-ground UAV flight altitude was adjusted the the changes in the field terrain. This resulted in a steady ground sampling distance (GSD) of the derived aerial images. In the field of weed mapping, this kind of altitude adjustments, using a digital terrain model, created from a preceding flight has never been tested before. The steady GSD of the aerial imagery enabled a precise size and height measurement of the target weeds through photogrammetric calculation of depth elevation models and orthomosaic images.

However, this study aimed to provide a tool for fast and robust weed mapping at field scale. By comparing the depth elevation models of the vegetation maps, application maps for *C. arvense* and *R. crispus* > 6 cm were calculated and resulted in herbicide savings of up to 90 %. After the first herbicide application in sugar beets, the total amount of *C. arvense* plants was reduced and further expansion was only observed at the edges of some distinct patches. This showed that the site-specific herbicide application was not only economically and ecologically reasonable, but also effective. The treatment of *C. arvense* in the four-leaf stage onward is also recommended by Tavaziva *et al.* (2019), at an average plant height of 6 cm. At this height, around the compensation point, where the acropetal allocation of carbohydrates turns into a basipetal movement of photosynthates (Nkurunziza and Streibig, 2011), *C. arvense* is highly susceptible to control measures, since the root system has the lowest regenerative capacity (Welton *et al.*, 1929). Thus, taking into account the plant height of *C. arvense* was an important step towards an ideal site-specific application, which has not been presented in earlier studies.

Fernandez-Quintanilla *et al.* (2018) report a generally low adoption of site-specific weed management systems by farmers, although some of these technologies have been available for several decades (Fountas *et al.*, 2005; Miller *et al.*, 2017). Hunt and Daughtry (2018) found that 60 % of the farmers in the USA have no access to variable rate ready application systems and cannot benefit from weed mapping. One reason for this is the system cost for the UAVs and sensors, which have to become cheaper to find more potential customers (Hunt and Daughtry, 2018). A second reason might be the known limitations in the classification of very small broad-leaved and grass weeds, also when ground-based mapping techniques were applied (Fernandez-Quintanilla *et al.*, 2018). However, even if the weed species cannot be identified, in most cases a differentiation can still be made between weeds and crops, with comparable cheap sensor systems (Burgos-Artizzu *et al.*, 2009). To meet the criteria for the use in practise, the experiments of the 1st study were conducted at field scale size. Hence, the flight plan settings, flight endurance and data collection were comparable to real field sizes at an average European farm and a compromise of area coverage and GSD was used for the UAV data collection. Therefore, weeds except for the size of *C. arvense* and *R. crispus* could not even be recognised on the orthomosaic image with the bare eye. Thus, no distinct weed shapes were visible, which could be used for an OBIA or pattern recognition approach. To result high-resolution aerial imagery, other studies often investigate smaller areas to create high-resolution datasets from single low altitude images for feature learning-based classification approaches (Hung *et al.*, 2014; Huang *et al.*, 2018).

The results of the presented research were in line with earlier findings of Peña *et al.* (2013), Peña *et al.* (2015), López-Granados *et al.* (2016b) and Fernandez-Quintanilla *et al.* (2018) for UAV weed identification in wide-row crops. The UAV-based aerial imagery weed mapping was not feasible in narrow-row cereals (Torres-Sánchez *et al.*, 2014), as well as the discrimination between multiple weed species or even monocotyledonous and dicotyledonous weeds was not possible.

As an alternative ground-based sensors for weed species discrimination at early vegetation stages can be used (Gerhards and Oebel, 2006; Weis and Gerhards, 2007; Andújar *et al.*, 2011a). Thereby, the ground vehicle-mounted cameras need an additional field passage for data collection, or real-time weed mapping, using a direct herbicide application have to be embedded on the same system (vehicle) (Gerhards and Christensen, 2003). However, if no direct-injection systems are available, real-time approaches may lead again to residual herbicides in the sprayer which limits the savings. Besides this, it has to be mentioned that both, aerial- and ground-based, mapping approaches are limited to post-emergence herbicide applications.

Like the other systems, also the UAV-based weed mapping has some limitations that are difficult to overcome. The sensors mounted on small and lightweight UAVs as used for field monitoring are limited in size and on the payload of the UAVs. Thus, flying at field scale with the currently available sensor resolutions will always be limited either in the ground sampling distance GSD or in the area coverage (Peña *et al.*, 2015). Nevertheless, the development of sensor techniques, UAV flight time, and payloads will be continued since UAVs provide many advantages compared to manned aerial vehicles and satellite imagery. Besides the high costs, manned aerial vehicles and even more satellites do not provide the high flexibility of UAVs and both are affected by the weather conditions (Hunt and Daughtry, 2018).

The developed algorithm showed a high accuracy in the classification of *C. arvense* and *R. crispus*, enabling site-specific herbicide applications. The herbicide saving potentials shown in this and the discussed studies should be the motivation to further improve the UAV image analysis for weed mapping. A special focus should, therefore, be placed on the use of RGB cameras with the highest possible resolution, for the detection of weeds at early vegetation states. In addition, the further development of suitable detection possibilities using multispectral fingerprints is promising if sensors with a sufficient resolution become available (Mesas-Carrascosa *et al.*, 2016).

The excessive use of agrochemicals has arisen public awareness and caused political action on the reduction of the chemical burden to the fields. Some maize herbicides became a selective weed control agent only by the additional application of a corresponding crop safener (Kraehmer *et al.*, 2014b). Nevertheless, this causes even higher chemical pollution to a field. The first commercially available safeners in the 1970s and 1980s were almost exclusively applied as soil-active compounds or directly as a seed treatment (Kraehmer *et al.*, 2014a). Nowadays, most safeners are exclusively sold as crop-selective ready-to-use mixtures with the herbicidal compounds. A possible safener reduction (*cyprosulfamide*) while maintaining the needed crop/weed selectivity of the herbicide (*isoxaflutole*) was tested in field and laboratory experiments in the 2nd study (Section 2.2). Through

UAV aerial and ground-based multispectral imagery, as well as hyperspectral sensor ground-based field monitoring of the experiments, the maize stress in the investigated *cyprosulfamide* treatments was quantified following an *isoxaflutole* treatment. The sensor-based results exceeded the significance of the auditor-based visual estimates. By applying the safener as a seed treatment, only 12 % of the application quantity was required compared to the commercial pre-mixture without significant yield losses.

Beside the examined reduction and application as seed treatment of *cyprosulf-amid* another aim was to supplement auditor-based visual estimates with spectral sensor measurements. By using spectral sensor-based experiment assessments, destructive measurements could be reduced to a minimum and additional inform-ation could be collected that is not visible to the human eye. The use of hyper-spectral sensors and multispectral imagery to measure the plant's reflectance of electromagnetic sunlight radiation enables plant stress assessments (Lichtenhaler *et al.*, 1998). Here, the calculation of vegetation indices (VIs) provides a suitable solution for a precise and comparable evaluation of plant stress (Lichtenhaler *et al.*, 1998). To detect the *isoxaflutole* caused plant damages, indices correlating with the chlorophyll content of leaves were chosen (Lichtenhaler *et al.*, 1998; Yancheva *et al.*, 2016).

The sensor data evaluation, as well as the other destructive and non-destructive measurements of the experiments, were in line with previous studies of Robinson *et al.* (2013), Philbrook and Santel (2008) and Ahrens *et al.* (2013). Although the reduction of the safener was possible, without significant yield losses, especially the UAV imagery of the field trial provided deep insights into the variability between the treatments and blocks. Classical data samples (e.g. biomass cuts or plant damage estimations) consist only of a few measurements, compared to sensor derived information, which is often more comprehensive and mostly collecting time series of information or multiple measurements of a probe. While the multiple measurements increase the raw data input at the statistical evaluation, in most cases it improves the significance of the results due to a higher selectivity (discriminatory power). Due to the great influence of data quality, particular attention must be laid on data collection during sensor measurement. In the presented study, the raw files of the hyperspectral sensor and the multispectral camera were used to calculate VIs. The aerial imagery of the field trial included a high number of non-target plants (*C. arvense*) since the focus was set on maize stress detection. Thus, the evaluation on the raw data including the patches with a dense soil coverage of *C. arvense* would have lead to misinterpretation of the maize stress. To avoid misclassification, the data was filtered from the raw image by the crop row position as already presented in the 1ˢᵗ study (Mink *et al.*, 2018). Thereby the selection has to be done very carefully in order not to lose any information on crop status. The same can be stated for the greenhouse pot trial, where the background noise in the images had to be filtered. Nevertheless, these images provided a lot more information than the auditor estimates, since the complete plant was measured and thus the stress origin or the differences to a previous image could be compared. In contrast the hyperspectral point spectrometer was measuring only one pixel, but with a 10 nm resolution from 360 to 1000 nm. This enabled plant stress detection prior to the appearance of visible damage. The daywise measurements of the hyper- and multispectral sensors, corrected by a standard white reference, enabled more significant statements than auditor estimates and could be endlessly repeated without affecting the target in contrast to destructive measurements.

Besides this, the sensor-based plant stress assessments also showed some limitations. The light dependency of the plant reflectance measurements often leads to high variability of the results (Thenkabail *et al.*, 2018b). Although other factors as nutrient shortage or plant age are also affecting the plant response, at homogeneous and repeated experiment conditions, the daywise reflectance differences were the most critical factor during the evaluation, which is also discussed by Peteinatos *et al.* (2016).

However, the correlation of the spectrometer VIs strongly correlated with the multispectral imagery derived VIs, which allowed a comparison of the sensor results. Additionally, both sensors were almost in line with the destructive measurements but gave deeper insights into treatments dependent plant responses. These were important findings since it allowed consistent quantitative statements on crop vitality after treatment and may also assist further research.

The saving potential of the *cyprosulfamide* safener seed treatment in *isoxaflutole* treated maize was found to be highly dependent on the weather conditions and maize plant development stage. Likewise, the UAV field monitoring indicated a high recovery potential in previously highly damaged plots, which was also found by Robinson *et al.* (2013). Thus, the effect of transient maize stress after *isoxaflutole* applications should be investigated. In further studies, focus should be put on the intra-maize specific competition eventually caused by photorespiration (de Veau and Burris, 1989; Ogren, 1975), nutrient, water and light competition (Maddonni and Otegui, 2004; Testa *et al.*, 2016), to separate herbicide stress from other stress origins. Regarding UAV and spectrometer hyper- and multispectral plant reflectance measurements for crop stress identification, further research has to be conducted on the day- and incident radiation dependency of the results in stressed crop plants. Although preceding studies have already shown the impact and possible solutions to avoid hyperspectral sensor and imagery interferences by changing incident light conditions (Makdessi *et al.*, 2017), sensor calibration and measuring at a fixed daytime is not sufficient to completely eliminate the effects. It should also be examined whether, in addition to the *4-hydroxyphenylpyruvate-dioxygenase* (HPPD)-inhibitor group, plant response to other herbicides can also be assessed with similar accuracy by leaf reflection measurements which may not cause concise leaf bleaching.

In the same way that herbicide-induced stress can be assessed in crops using a chlorphyll fluorometer (Weber *et al.*, 2017) or by measuring the hyperspectral plant reflectance (Thenkabail *et al.*, 2018b), it can be also be recorded in weeds (Kaiser *et al.*, 2013; Reddy *et al.*, 2014). Herbicide stress assessments in crops mainly focus on the selectivity and compatibility of herbicide treatments. In the case of weeds, herbicide sensitivity measurements aim to identify weeds suspected to be herbicide-resistant. With the growing number of herbicide-resistant weeds, the need for reliable test methods increased and screening protocols have been established (Beckie *et al.*, 2000; Burgos *et al.*, 2013; Beckie and Harker, 2017). These screening methods are mostly based on pot bioassays, using viable seeds collected from suspicious populations at the fields (Burgos, 2015). Bioassays provide reliable results on herbicide-resistance (Streibig, 1988). Nevertheless, the duration of these screenings may easily exceed two months, virtually excluding targeted in-season control of resistant weeds. Therefore, the 3rd and 4th study (Sections 2.3 and 2.4) aimed at the development of an in-field, in-season quick test for herbicide-resistance identification of individual plants.

The 3rd study employed the assessment of chlorophyll fluorescence parameters in the target weeds using a pulse amplitude modulated (PAM)-imaging chlorophyll fluorometer for the herbicide-resistance classification. The maximum quantum efficiency of photosystem II (F_v/F_m) is commonly used in plant physiology to assess photosynthetic efficiency (Baker, 2008). In 2013, Kaiser *et al.* (2013) presented the identification of *acetolactate* synthase (ALS) and *acetyl-CoA carboxylase* (ACCase) resistant *Alopecurus myosuroides* Huds. populations by the F_v/F_m ratio under laboratory conditions. Three years later Wang *et al.* (2016) evaluated the in-field F_v/F_m assessment in *A. myosuroides* using a mobile PAM-imaging chlorophyll fluorometer prototype. Meanwhile, many studies presented herbicide-resistance monitoring in monocotyledonous weeds using the F_v/F_m for ACCase-, ALS-, and photosystem II-inhibiting herbicides (Wang *et al.*, 2016; Kaiser *et al.*, 2013; Wang *et al.*, 2017; Linn *et al.*, 2018). The F_v/F_m based detection of ALS-inhibitor herbicide-resistant dicotyledonous weeds was first presented in the 3rd study of this dissertation by Linn *et al.* (2019).

In this study, the F_v/F_m ratios of herbicide sensitive and resistant *Papaver rhoeas* and *Stellaria media* plants were examined after the application of the ALS-inhibitors *metsulfuron-methyl, tribenuron-methyl,* and *florasulam*. Already three days after treatment with the herbicides (DAT), the sensitive *P. rhoeas* and *S. media* plants showed a significant decrease of 28 % and 48 % in the F_v/F_m ratios, respectively, compared to the untreated control. However, the resistant *P. rhoeas* and *S. media* plants indicated no decrease in the F_v/F_m ratios. This allowed a statistical differentiation of herbicide-resistant and -sensitive populations in both species by the mean F_v/F_m values.

Since the F_v/F_m-based statistical evaluation may not always allow a precise resistance classification, if an initial decrease of the F_v/F_m ratio is observed, classifiers have been trained. The classifiers enabled a precise classification to the classes "susceptible" or "resistant", by a maximum-likelihood classification of the F_v/F_m ratios, observed in the sensitive and resistant plants of the investigated populations. The classification accuracies at 3 DAT reached up to 100 % and 95 % for *S. media* and *P. rhoeas*, respectively. These results showed that the identification of resistant plants within a population was possible. Depending on the accuracy of the classification results, an estimation of the frequency of resistant plants within the population is possible.

Nevertheless, the field assessments of the F_v/F_m values is rather time intensive. The dark adoption of the plants, prior to the measurements will prevent a fast data collection at large areas by the PAM-imaging sensor. Despite this limitation, the presented classification procedure significantly accelerated and simplified previous assessments. Although there are slight decreases in the classification accuracy of the resistant *P. rhoeas* population, especially for *florasulam*, the significance of the result still exceeds Tukey's HSD test. Thus, direct in-season weed control of herbicide-resistant plants is possible.

The field trials conducted for the 3rd and 4th study, were located at two different trial stations at slightly different climatic conditions and on different soils. Yet, the plants for the first vegetation period were pre-grown in the greenhouse and herbicide was applied in the laboratory track sprayer, prior to the field transplantation, to ensure a uniform herbicide application. In the experiment, these factors enabled an ideal situation at the trial start using identical treated plants. However, both classifiers might have failed the validation on real field plants, if trained with the uniformly grown and treated plants from the experiment. Under field conditions many factors are influencing the plant response to measurements. In this case the chlorophyll fluorescence measurement might exceed the hyperspectral measurements, due to the measurement of the F_v/F_m ratio, while the plant reflectance changes its spectral intensity at maturing already in the visible spectral range (Burks *et al.*, 2002).

A second approach for the classification of herbicide-resistant *S. media* plants was realised in the same field experiments as presented in the 3rd study (Section 2.3), but with an additional vegetation period. The procedure and results of this approach are described in the 4th study (Section 2.4). As reviewed in Stafford (2000), "Precision agriculture is 'information-intense' and could not be realised without the enormous advances in networking and computer processing power." The link between data connections and computing power becomes most visible in the use of an artificial neural network (ANN) for training and classification purposes. An unfiltered dataset of the ALS-inhibitor sensitive and resistant *S. media* plants, assessed with a spectrometer, was used to train an ANN to the two output classes "resistant" and "sensitive". The use of hyperspectral sensors for plant stress evaluation is a widely used tool in daily farming practise and science. Many studies are using hyperspectral either derived from satellites (Soria-Ruíz and Fernández-Ordoñez, 2003) or UAV (Zaman-Allah *et al.*, 2015) for detailed information on crop vigour at field scale. Ground-based assessments of plant stress is also reviewed in Peteinatos *et al.* (2016) using multiple sensors including hyperspectral plant reflectance. Peteinatos *et al.* (2016) also described the use of research objective related VIs, of which the NDVI is the best-known.

The measurement of hyperspectral plant reflectance and evaluation of herbicide-resistant weeds was already presented by Reddy *et al.* (2014) and Matzrafi *et al.* (2017) in *glyphosate*-resistant *Amaranthus palmeri* and by Lee *et al.* (2014) in *glyphosate*-resistant *Lolium multiflorum* plants. In these studies, the evaluation of the herbicide sensitivity of the populations was not VI based, but classifiers were trained on raw data. However, these studies used subdivided datasets for their training and classification. Thus, the 4th study of this work (Section 2.4) used a passive sensor without an artificial light source to measure herbicide-resistant and sensitive *S. media* plants at two locations, with the first location at a second vegetation period. This way the ANN could be trained with the unfiltered raw data from a completely independent dataset from one location.

The ANN achieved a classification accuracy of up to 95 % at 4 DAT in the first vegetation period and still 72 % in the following year. Thus, the classifier could identify sensitive plants before any visual symptoms appeared. The following analysis of the neuron weights, which evolved in the ANN during the training process provided information on the importance of the 65 input spectra. This was proved by the training of a second ANN using the 12 most important spectra, and a third ANN with 12 less important spectra. The classification accuracy of the second ANN showed similar or improved results compared to the full spectra ANN, while the results of the third ANN weakened.

The findings were in line with the previously presented study of Linn *et al.* (2019), where similar classification results were found. The classification accuracies of the ANNs using the unfiltered rawdata of the spectrometer for the training demonstrated the power of ANNs to elicit reliable conclusions from complex and tremendous amounts of data (Zhang *et al.*, 1998). Here, the ANN showed a better performance than the hyperspectral plant reflectance classifier presented by Reddy *et al.* (2014), where an artificial light source was used and the training was performed with a subset of the original data. The improved classification accuracies at a reduced spectra input was in line with the findings of Reddy *et al.* (2014). These findings can be explained by the Huges phenomenon (Hughes, 1968), which states, that the benefit of adding more input is outweighed at some point by the increased error in statistical parameter estimates.

However, it has to be mentioned that the decreasing accuracies of the ANNs in the second vegetation period have to be improved if the application has to fulfil the requirements for a reliable herbicide-resistance test. Again the measurement method has its limits if a point spectrometer is used. Even though we conducted five measurement repetitions per plant, the portion of mismeasurements was relatively high, due to a very short sensor to plant distance of 5 cm. This often leads to a high soil portion in the measurement, when measuring the comparatively small-sized *S. media* plants. In addition, it had to be ensured that at the short measurement distance, the sensor did not influence the incident light and thus could have shaded parts of the plant. From this point of view, the classification accuracy was strong, yet the measurement method should be improved in further studies.

While a reduced lens angle of the spectrometer could help to avoid these mixed pixels (soil/plant) effects, a more sophisticated solution would be the use of a multi- or hyperspectral camera. The channels needed in such a camera can be found in the 12 spectra of the second ANN. If the data is provided as an image, also the training of a convolutional neural network could be realised, which are known to work best on images and have already outperformed the fully connected neural networks as used in our trial (Singh *et al.*, 2018). Further experiments should focus on a pre-filtering, to evaluate if the training with unfiltered raw data has also a negative side aspect in the classification accuracy. Additionally, other weed species, including monocotyledon weeds should be investigated, even if the increasing problem of the spectrometer's recording area must be taken into account.

The 4th study demonstrated the suitability of the passive hyperspectral sensor as a suitable tool for herbicide-resistance detection. Thus, with the help of this sensor in conjunction with the ANN, herbicide-resistance can be detected at an early stage and control measures can be adapted.

Overall the studies demonstrated the potential of PF applications to reduce herbicide inputs in maize and sugar beet, adopt chemical weed management strategies to avoid maize stress, and detect herbicide-resistance in *P. rhoeas* and *S. media*. Therefore, the studies have contributed to the further development of UAV and sensor-based PF applications, encouraging further research and commercialisation. The results of the tested sensor applications have proven to be a complementary and sometimes even superior research tool to auditor estimates. Besides the promising results, the reported issues have to be tackled to bring the applications into use in farming systems and to reduce the negative impact of chemical plant protection to a minimum.

Thus future research on the presented topic should further focus on artificial neural networks (e.g. convolutional neural networks) image analysis for the weed species detection on aerial imagery. In any case the GSD must be adopted to the smallest objects to be detected or identified (Sandau, 2005). Sandau (2005) desribe the GSD for the object dectetion to object size divided by 3 and an even more ground pixels for the object identification. To further improve the potentials of UAVs to be used as small and lightweight sensor carriers, which can be easily transported and operate even on relatively small survey areas, some limitations have to be further improved. The two major UAV system limitations concerning multirotor platforms are the limited flight endurance and the need for an operator to survey, command and control the flight. Here a technical and legal basis has to enable flights beyond the visual line of sight (BVLOS) for future UAV use cases in farming practise.

Bibliography

Ahrens, H., Lange, G., Müller, T., Rosinger, C., Willms, L., and van Almsick, A. (2013). 4-hydroxyphenylpyruvate dioxygenase inhibitors in combination with safeners: solutions for modern and sustainable agriculture. *Angewandte Chemie International Edition*, 52(36):9388–9398.

Ahrens, W. H., Arntzen, C. J., and Stoller, E. W. (1981). Chlorophyll fluorescence assay for the determination of triazine resistance. *Weed Science*, 29(3):316–322.

Andújar, D., Escolà, A., Dorado, J., and Fernández-Quintanilla, C. (2011a). Weed discrimination using ultrasonic sensors. *Weed Research*, 51(6):543–547.

Andújar, D., Ribeiro, A., Carmona, R., Fernández-Quintanilla, C., and Dorado, J. (2010). An assessment of the accuracy and consistency of human perception of weed cover. *Weed Research*, 50(6):638–647.

Andújar, D., Ribeiro, A., Fernández-Quintanilla, C., and Dorado, J. (2011b). Accuracy and feasibility of optoelectronic sensors for weed mapping in wide row crops. *Sensors*, 11(3):2304–2318.

Baker, N. R. (2008). Chlorophyll fluorescence: a probe of photosynthesis in vivo. *Annual Review of Plant Biology*, 59(1):89–113.

Beckie, H. J. and Harker, K. N. (2017). Our top 10 herbicide-resistant weed management practices. *Pest Management Science*, 73(6):1045–1052.

Beckie, H. J., Heap, I. M., Smeda, R. J., and Hall, L. M. (2000). Screening for Herbicide Resistance in Weeds. *Weed Technology*, 14(2):428–445.

Bivand, R., Keitt, T., Rowlingson, B., Pebesma, E., Sumner, M., Hijmans, R., and Rouault, E. (2017). rgdal: Bindings for the 'Geospatial' Daa Abstraction Library. (https://cran.r-project.org/web/packages/rgdal/).

Borra-Serrano, I., Peña, J. M., Torres-Sánchez, J., Mesas-Carrascosa, F., and López-Granados, F. (2015). Spatial Quality Evaluation of Resampled Unmanned Aerial Vehicle-Imagery for Weed Mapping. *Sensors*, 15(8):19688–19708.

Burgos, N. R. (2015). Whole-plant and seed bioassays for resistance confirmation. *Weed Science*, 63(sp1):152–165.

Burgos, N. R., Tranel, P. J., Streibig, J. C., Davis, V. M., Shaner, D., Norsworthy, J. K., and Ritz, C. (2013). Review: Confirmation of Resistance to Herbicides and Evaluation of Resistance Levels. *Weed Science*, 61(1):4–20.

Burgos-Artizzu, X. P., Ribeiro, A., Tellaeche, A., Pajares, G., and Fernández-Quintanilla, C. (2009). Improving weed pressure assessment using digital images from an experience-based reasoning approach. *Computers and Electronics in Agriculture*, 65(2):176–185.

Burks, T. F., Shearer, S. A., Green, J. D., and Heath, J. R. (2002). Influence of weed maturity levels on species classification using machine vision. *Weed Science*, 50(6):802–811.

Carter, G. A. (1998). Reflectance wavebands and indices for remote estimation of photosynthesis and stomatal conductance in pine canopies. *Remote Sensing of Environment*, 63(1):61–72.

Chantre, G. R., Vigna, M. R., Renzi, J. P., and Blanco, A. M. (2018). A flexible and practical approach for real-time weed emergence prediction based on Artificial Neural Networks. *Biosystems Engineering*, 170:51–60.

Chen, Y., Lin, Z., Zhao, X., Wang, G., and Gu, Y. (2014). Deep Learning-Based Classification of Hyperspectral Data. *IEEE Journal of Selected Topics in Applied Earth Observations and Remote Sensing*, 7(6):2094–2107.

Christensen, S., Rasmussen, J., Pedersen, S. M., Dorado, J., and Fernández-Quintanilla, C. (2014). Prospects for Site Specific Weed Management. In Pablo Gonzales-de Santos, A. R., editor, *International Conference on Robotics and Associated High-Technologies and Equipment for Agriculture and Forestry*, number May, pages 541–549, Madrid, Spain.

Christensen, S., Søgaard, H. T., Kudsk, P., Nørremark, M., Lund, I., Nadimi, E. S., and Jørgensen, R. (2009). Site-specific weed control technologies. *Weed Research*, 49(3):233–241.

Core Team R (2018). R: A language and environment for statistical computing. *R Foundation for Statistical Computing*, (https://www.r-project.org/).

Cousens, R. (1985). A simple model relating yield loss to weed density. *Annals of Applied Biology*, 107(2):239–252.

Dan Hess, F. (2018). Herbicide absorption and translocation and their relationship to plant tolerances and susceptibility. In *Weed Physiology*, pages 191–214. CRC Press.

Davies, J. and Caseley, J. C. (1999). Herbicide safeners: a review. *Pesticide Science*, 55(11):1043–1058.

de Castro, A. I., Torres-Sánchez, J., Peña, J., Jiménez-Brenes, F., Csillik, O., and López-Granados, F. (2018). An Automatic Random Forest-OBIA Algorithm for Early Weed Mapping between and within Crop Rows Using UAV Imagery. *Remote Sensing*, 10(3):285.

de Veau, E. J. and Burris, J. E. (1989). Photorespiratory rates in wheat and maize as determined by o-labeling. *Plant physiology*, 90(2):500–11.

Eitel, J., Keefe, R., Long, D., Davis, A., and Vierling, L. (2010). Active ground optical remote sensing for improved monitoring of seedling stress in nurseries. *Sensors*, 10(4):2843–2850.

Eitel, J. U., Vierling, L. A., Litvak, M. E., Long, D. S., Schulthess, U., Ager, A. A., Krofcheck, D. J., and Stoscheck, L. (2011). Broadband, red-edge information from satellites improves early stress detection in a New Mexico conifer woodland. *Remote Sensing of Environment*, 115(12):3640–3646.

Fernández-Quintanilla, C., Peña-Barragán, J. M., Andújar, D., Dorado, J., Ribeiro, A., and López-Granados, F. (2018). Is the current state-of-the-art of weed monitoring suitable for site-specific weed management in arable crops? *Weed Research*, 58(4):259–272.

Fountas, S., Blackmore, S., Ess, D., Hawkins, S., Blumhoff, G., Lowenberg-Deboer, J., and Sorensen, C. G. (2005). Farmer experience with precision agriculture in denmark and the US eastern corn belt. *Precision Agriculture*, 6(2):121–141.

Geipel, J. and Korsaeth, A. (2017). Hyperspectral Aerial Imaging for Grassland Yield Estimation. *Advances in Animal Biosciences*, 8(2):770–775.

Geipel, J., Link, J., and Claupein, W. (2014). Combined Spectral and Spatial Modeling of Corn Yield Based on Aerial Images and Crop Surface Models Acquired with an Unmanned Aircraft System. *Remote Sensing*, 6(11):10335–10355.

Gerhards, R. (2010). Spatial and Temporal Dynamics of Weed Populations. In Oerke, E. C., Gerhards, R., Menz, G., and Sikora, R., editors, *Precision Crop Protection - the Challenge and Use of Heterogeneity*, pages 17–25. Springer, Netherlands, Dordrecht.

Gerhards, R. and Christensen, S. (2003). Real-time weed detection, decision making and patch spraying in maize, sugarbeet, winter wheat and winter barley. *Weed Research*, 43(6):385–392.

Gerhards, R. and Oebel, H. (2006). Practical experiences with a system for site-specific weed control in arable crops using real-time image analysis and GPS-controlled patch spraying. *Weed Research*, 46(3):185–193.

Gerhards, R., Sökefeld, M., Knuf, D., and Kühbauch, W. (1996). Kartierung und geostatistische Analyse der Unkrautverteilung in Zuckerrübenschlägen als Grundlage für eine teilschlagspezifische Bekämpfung. *J. Agron. & Crop Science*, 176:259–266.

Gerhards, R., Sökefeld, M., Timmermann, C., Kühbauch, W., and Williams II, M. M. (2002). Site-Specific Weed Control in Maize, Sugar Beet, Winter Wheat, and Winter Barley. *Precision Agriculture*, 3(1):25–35.

Gitelson, A. A., Kaufman, Y. J., and Merzlyak, M. N. (1996). Use of a green channel in remote sensing of global vegetation from EOS-MODIS. *Remote Sensing of Environment*, 58(3):289 – 298.

Goodfellow, I., Bengio, Y., and Courville, A. (2016). *Deep Learning*. The MIT Press, Cambridge, Massachusetts, USA.

Gorman, R. and Sejnowski, T. J. (1988). Analysis of hidden units in a layered network trained to classify sonar targets. *Neural Networks*, 1(1):75–89.

Gutjahr, C. and Gerhards, R. (2010). Decision Rules for Site-Specific Weed Management. In *Precision Crop Protection - the Challenge and Use of Heterogeneity*, pages 223–239. Springer Netherlands, Dordrecht.

Gutjahr, C., Sökefeld, M., and Gerhards, R. (2012). Evaluation of two patch spraying systems in winter wheat and maize. *Weed Research*, 52(6):510–519.

Håkansson, S. (2003). *Weeds and weed management on arable land: an ecological approach*. CABI, Wallingford, United Kingdom.

Hall, M. R., Swanton, C. J., and Anderson, G. W. (1992). The critical period of weed control in grain corn (Zea mays). *Weed Science*, 40(3):441–447.

Hamouz, P., Hamouzová, K., Holec, J., and Tyšer, L. (2018). Impact of site-specific weed management on herbicide savings and winter wheat yield. *Plant, Soil and Environment*, 59(No. 3):101–107.

Hamuda, E., Glavin, M., and Jones, E. (2016). A survey of image processing techniques for plant extraction and segmentation in the field. *Computers and Electronics in Agriculture*, 125:184–199.

Hasanuzzaman, M., Fujita, M., Oku, H., Nahar, K., and Hawrylak-Nowak, B. (2018). *Plant nutrients and abiotic stress tolerance*. Springer Singapore.

Hatzios, K. K. and Hoagland, R. E. (1989). *Crop safeners for herbicides: development, uses, and mechanisms of action*. Academic Press, San Diego, USA.

Heap, I. (2019). The International Survey of Herbicide Resistant Weeds. *http://www.weedscience.org, last accessed 20.08.2019*.

Heege, H. J., editor (2013). *Precision in Crop Farming*. Springer Netherlands, Dordrecht.

Hijmans, R. J., Etten, J. v., Cheng, J., Mattiuzzi, M., Sumner, M., Greenberg, J. A., Lamigueiro, O. P., Bevan, A., Racine, E. B., Shortridge, A., and Ghosh, A. (2017). raster: Geographic analysis and modeling with raster data. (https://cran.r-project.org/web/packages/raster/raster.pdf).

Hock, B., Fedtke, C., and Schmidt, R. (1995). *Herbizide*. Thieme, Stuttgart, Germany.

Huang, H., Deng, J., Lan, Y., Yang, A., Deng, X., and Zhang, L. (2018). A fully convolutional network for weed mapping of unmanned aerial vehicle (UAV) imagery. *PLOS ONE*, 13(4):e0196302.

Hughes, G. (1968). On the mean accuracy of statistical pattern recognizers. *IEEE Transactions on Information Theory*, 14(1):55–63.

Hung, C., Xu, Z., and Sukkarieh, S. (2014). Feature Learning Based Approach for Weed Classification Using High Resolution Aerial Images from a Digital Camera Mounted on a UAV. *Remote Sensing*, 6(12):12037–12054.

Hunt, E. R. and Daughtry, C. S. T. (2018). What good are unmanned aircraft systems for agricultural remote sensing and precision agriculture? *International Journal of Remote Sensing*, 39(15-16):5345–5376.

International Seed Testing Association (2019). *International rules for seed testing.* Number 1. (https://www.seedtest.org).

Jensen, H. G., Jacobsen, L.-B., Pedersen, S. M., and Tavella, E. (2012). Socioeconomic impact of widespread adoption of precision farming and controlled traffic systems in Denmark. *Precision Agriculture*, 13(6):661–677.

Johnson, G. A., Mortensen, D. A., and Martin, A. R. (1995). A simulation of herbicide use based on weed spatial distribution. *Weed Research*, 35(3):197–205.

Johnson, W. G., Bradley, P. R., Hart, S. E., Buesinger, M. L., and Massey, R. E. (2000). Efficacy and economics of weed management in glyphosate-resistant corn (Zea mays). *Weed Technology*, 14(1):57–65.

Kaiser, Y. I., Menegat, A., and Gerhards, R. (2013). Chlorophyll fluorescence imaging: a new method for rapid detection of herbicide resistance in Alopecurus myosuroides. *Weed Research*, 53(6):399–406.

Kamilaris, A. and Prenafeta-Boldú, F. X. (2018). Deep learning in agriculture: A survey. *Computers and Electronics in Agriculture*, 147:70–90.

Keller, M., Gutjahr, C., Möhring, J., Weis, M., Sökefeld, M., and Gerhards, R. (2014). Estimating economic thresholds for site-specific weed control using manual weed counts and sensor technology: An example based on three winter wheat trials. *Pest Management Science*, 70(2):200–211.

Keränen, M., Aro, E.-M., Tyystjärvi, E., and Nevalainen, O. (2003). Automatic Plant Identification with Chlorophyll Fluorescence Fingerprinting. *Precision Agriculture*, 4(1):53–67.

Knezevic, S. Z., Sikkema, P. H., Tardif, F., Hamill, A. S., Chandler, K., and Swanton, C. J. (1998). Biologically effective dose and selectivity of RPA 201772 for preemergence weed control in corn (Zea mays). *Weed Technology*, 12(4):670–676.

Kraehmer, H., Laber, B., Rosinger, C., and Schulz, A. (2014a). Herbicides as weed control agents: state of the art: I. Weed control research and safener technology: The path to modern agriculture. *Plant physiology*, v. 166(no. 3):1119–1131.

Kraehmer, H., van Almsick, A., Beffa, R., Dietrich, H., Eckes, P., Hacker, E., Hain, R., Strek, H. J., Stuebler, H., and Willms, L. (2014b). Herbicides as weed control agents: State of the art: II. Recent achievements. *Plant physiology*, 166(3):1132–1148.

Kudsk, P. and Streibig, J. C. (2003). Herbicides - a two-edged sword*. *Weed Research*, 43(2):90–102.

Lamastus, F. E. and Shaw, D. R. (2005). Comparison of different sampling scales to estimate weed populations in three soybean fields. *Precision Agriculture*, 6(3):271–280.

Lambert, J. P. T., Hicks, H. L., Childs, D. Z., and Freckleton, R. P. (2018). Evaluating the potential of Unmanned Aerial Systems for mapping weeds at field scales: a case study with Alopecurus myosuroides. *Weed Research*, 58(1):35–45.

Lee, M. A., Huang, Y., Nandula, V. K., and Reddy, K. N. (2014). Differentiating glyphosate-resistant and glyphosate-sensitive Italian ryegrass using hyperspectral imagery. volume 9108. International Society for Optics and Photonics.

Lichtenhaler, H. K., Wenzel, O., Buschmann, C., and Gittelson, A. (1998). Plant stress detection by reflectance and fluorescence. *Annals of the New York Academy of Sciences*, 851(Stress of Life: From Molecules to Man):271–285.

Liebisch, F., Kirchgessner, N., Schneider, D., Walter, A., and Hund, A. (2015). Remote, aerial phenotyping of maize traits with a mobile multi-sensor approach. *Plant Methods*, 11(1):9.

Lindblom, J., Lundström, C., Ljung, M., and Jonsson, A. (2017). Promoting sustainable intensification in precision agriculture: review of decision support systems development and strategies. *Precision Agriculture*, 18(3):309–331.

Linn, A. I., Košnarová, P., Soukup, J., and Gerhards, R. (2018). Detecting herbicide-resistant Apera spica-venti with a chlorophyll fluorescence agar test. *Plant, Soil and Environment*, 64(No. 8):386–392.

Linn, A. I., Mink, R., Peteinatos, G. G., and Gerhards, R. (2019). In-field classification of herbicide-resistant Papaver rhoeas and Stellaria media using an imaging sensor of the maximum quantum efficiency of photosystem II. *Weed Research*, page wre.12374.

Longchamps, L., Panneton, B., Simard, M.-J., and Leroux, G. D. (2012). Could Weed Sensing in Corn Interrows Result in Efficient Weed Control? *Weed Technology*, 26(04):649–656.

Longchamps, L., Panneton, B., Simard, M.-J., and Leroux, G. D. (2014). An Imagery-Based Weed Cover Threshold Established Using Expert Knowledge. *Weed Science*, 62(1):177–185.

López-Granados, F., Torres-Sánchez, J., de Castro, A. I., Serrano-Pérez, A., Mesas-Carrascosa, F. J., and Peña, J. M. (2016a). Object-based early monitoring of a grass weed in a grass crop using high resolution UAV imagery. *Agronomy for Sustainable Development*, 36(4):1–12.

López-Granados, F., Torres-Sánchez, J., Serrano-Pérez, A., de Castro, A. I., Mesas-Carrascosa, F. J., and Peña, J. M. (2016b). Early season weed mapping in sunflower using UAV technology: variability of herbicide treatment maps against weed thresholds. *Precision Agriculture*, 17(2):183–199.

López-Lozano, R., Casterad, M., and Herrero, J. (2010). Site-specific management units in a commercial maize plot delineated using very high resolution remote sensing and soil properties mapping. *Computers and Electronics in Agriculture*, 73(2):219–229.

Maccioni, A., Agati, G., and Mazzinghi, P. (2001). New vegetation indices for remote measurement of chlorophylls based on leaf directional reflectance spectra. *Journal of Photochemistry and Photobiology B: Biology*, 61(1-2):52–61.

Maddonni, G. and Otegui, M. (2004). Intra-specific competition in maize: early establishment of hierarchies among plants affects final kernel set. *Field Crops Research*, 85(1):1–13.

Mahlein, A.-K., Kuska, M. T., Behmann, J., Polder, G., and Walter, A. (2018). Hyperspectral Sensors and Imaging Technologies in Phytopathology: State of the Art. *Annual Review of Phytopathology*, 56(1):535–558.

Mahlein, A.-K., Steiner, U., Hillnhütter, C., Dehne, H.-W., and Oerke, E.-C. (2012). Hyperspectral imaging for small-scale analysis of symptoms caused by different sugar beet diseases. *Plant Methods*, 8(1):3.

Makdessi, N. A., Jean, P.-A., Ecarnot, M., Gorretta, N., Rabatel, G., and Roumet, P. (2017). How plant structure impacts the biochemical leaf traits assessment from in-field hyperspectral images: A simulation study based on light propagation modeling in 3D virtual wheat scenes. *Field Crops Research*, 205:95–105.

Mansourian, S., Darbandi, E. I., Rashed Mohassel, M. H., Rastgoo, M., and Kanouni, H. (2017). Comparison of artificial neural networks and logistic regression as potential methods for predicting weed populations on dryland chickpea and winter wheat fields of Kurdistan province, Iran. *Crop Protection*, 93:43–51.

Marshall, E. J. P. (1988). Field-scale estimates of grass weed populations in arable land. *Weed Research*, 28(3):191–198.

Marshall, R., Hull, R., and Moss, S. R. (2010). Target site resistance to ALS inhibiting herbicides in Papaver rhoeas and Stellaria media biotypes from the UK. *Weed Research*, 50(6):621–630.

Matzrafi, M., Herrmann, I., Nansen, C., Kliper, T., Zait, Y., Ignat, T., Siso, D., Rubin, B., Karnieli, A., and Eizenberg, H. (2017). Hyperspectral Technologies for Assessing Seed Germination and Trifloxysulfuron-methyl Response in Amaranthus palmeri (Palmer Amaranth). *Frontiers in Plant Science*, 8:474.

Mayonado, D. J., Hatzios, K. K., Orcutt, D. M., and Wilson, H. P. (1989). Evaluation of the mechanism of action of the bleaching herbicide SC-0051 by HPLC analysis. *Pesticide Biochemistry and Physiology*, 35(2):138–145.

Mechant, E., De Marez, T., Aper, J., and Bulcke, R. (2010). Chlorophyll fluorescence protocol for quick detection of triazinone resistant Chenopodium album L. *Communications in agricultural and applied biological sciences*, 75(2):83–90.

Mesas-Carrascosa, F. J., Clavero Rumbao, I., Torres-Sánchez, J., García-Ferrer, A., Peña, J. M., and López Granados, F. (2016). Accurate ortho-mosaicked six-band multispectral UAV images as affected by mission planning for precision agriculture proposes. *International Journal of Remote Sensing*, pages 1–16.

Mesas-Carrascosa, F.-J., Torres-Sánchez, J., Clavero-Rumbao, I., García-Ferrer, A., Peña, J. M., Borra-Serrano, I., and López-Granados, F. (2015). Assessing Optimal Flight Parameters for Generating Accurate Multispectral Orthomosaicks by UAV to Support Site-Specific Crop Management. *Remote Sensing*, 7(10):12793–12814.

Meyer, G. E., Hindman, T. W., and Laksmi, K. (1999). Machine vision detection parameters for plant species identification. volume 3543, pages 327–335. International Society for Optics and Photonics.

Meyer, G. E., Neto, J. C., Jones, D. D., and Hindman, T. W. (2004). Intensified fuzzy clusters for classifying plant, soil, and residue regions of interest from color images. *Computers and Electronics in Agriculture*, 42(3):161–180.

Miller, N. J., Griffin, T. W., Bergtold, J., Ciampitti, I. A., and Sharda, A. (2017). Farmers' Adoption Path of Precision Agriculture Technology. *Advances in Animal Biosciences*, 8(2):708–712.

Mink, R., Dutta, A., Peteinatos, G., Sökefeld, M., Engels, J., Hahn, M., and Gerhards, R. (2018). Multi-temporal site-specific weed control of Cirsium arvense (L.) Scop. and Rumex crispus L. in maize and sugar beet using unmanned aerial vehicle based mapping. *Agriculture*, 8(5):65.

Mortensen, D. A. and Coble, H. D. (1991). Two Approaches to Weed Control Decision-Aid Software. *Weed Technology*, 5(2):445–452.

Nkurunziza, L. and Streibig, J. C. (2011). Carbohydrate dynamics in roots and rhizomes of Cirsium arvense and Tussilago farfara. *Weed Research*, 51(5):461–468.

Norsworthy, J. K., Talbert, R. E., and Hoagland, R. E. (1999). Chlorophyll fluorescence evaluation of agrochemical interactions with propanil on propanil-resistant barnyardgrass (Echinochloa crus-galli). *Weed Science*, 47(1):13–19.

Nugent, P. W., Shaw, J. A., Jha, P., Scherrer, B., Donelick, A., and Kumar, V. (2018). Discrimination of herbicide-resistant kochia with hyperspectral imaging. *Journal of Applied Remote Sensing*, 12(01):1.

Ogren, W. L. (1975). Control of Photorespiration in Soybean and Maize. In *Environmental and Biological Control of Photosynthesis*, pages 45–52. Springer Netherlands, Dordrecht.

Owen, M. D. and Zelaya, I. A. (2005). Herbicide-resistant crops and weed resistance to herbicides. *Pest Management Science*, 61(3):301–311.

Pallett, K., Little, J., Sheekey, M., and Veerasekaran, P. (1998). The Mode of Action of Isoxaflutole: I. Physiological Effects, Metabolism, and Selectivity. *Pesticide Biochemistry and Physiology*, 62(2):113–124.

Peña, J. M., Torres-Sánchez, J., de Castro, A. I., Kelly, M., and López-Granados, F. (2013). Weed Mapping in Early-Season Maize Fields Using Object-Based Analysis of Unmanned Aerial Vehicle (UAV) Images. *PLoS ONE*, 8(10):e77151.

Peña, J. M., Torres-Sánchez, J., Serrano-Pérez, A., de Castro, A. I., and López-Granados, F. (2015). Quantifying Efficacy and Limits of Unmanned Aerial Vehicle (UAV) Technology for Weed Seedling Detection as Affected by Sensor Resolution. *Sensors*, 15(3):5609–5626.

Peñuelas, J. and Filella, I. (1998). Visible and near-infrared reflectance techniques for diagnosing plant physiological status. *Trends in Plant Science*, 3(4):151–156.

Pérez-Ortiz, M., Peña, J. M., Gutiérrez, P. A., Torres-Sánchez, J., Hervás-Martínez, C., and López-Granados, F. (2016). Selecting patterns and features for between- and within- crop-row weed mapping using UAV-imagery. *Expert Systems with Applications*, 47:85–94.

Peteinatos, G. G., Korsaeth, A., Berge, T. W., and Gerhards, R. (2016). Using optical sensors to identify water deprivation, nitrogen shortage, weed presence and fungal infection in wheat. *Agriculture*, 6(2):24.

Peteinatos, G. G., Weis, M., Andújar, D., Rueda Ayala, V., and Gerhards, R. (2014). Potential use of ground-based sensor technologies for weed detection. *Pest Management Science*, 70(2):190–199.

Philbrook, B. D. and Santel, H. J. (2008). A new formulation of Isoxaflutole for preemergence weed control in corn (Zea mays). In *Weed Science Society of America Meeting*.

Rasmussen, J., Nielsen, J., Garcia-Ruiz, F., Christensen, S., and Streibig, J. C. (2013). Potential uses of small unmanned aircraft systems (UAS) in weed research. *Weed Research*, 53(4):242–248.

Rasooli Sharabiani, V., Noguchi, N., Han-Ya, I., and Ishi, K. (2013). Evaluation of an active remote sensor for monitoring winter wheat growth status. *Engineering in Agriculture, Environment and Food*, 6:118–127.

Rassmussen, J. (1991). A model for prediction of yield response in weed harrowing. *Weed Research*, 31(6):401–408.

Reckleben, Y. (2014). Sensors for nitrogen fertilization - Experiences of 12 years practical use. *Journal für Kulturpflanzen*, 66(2):42–47.

Reddy, K. N., Huang, Y., Lee, M. A., Nandula, V. K., Fletcher, R. S., Thomson, S. J., and Zhao, F. (2014). Glyphosate-resistant and glyphosate-susceptible Palmer amaranth (Amaranthus palmeri S. Wats.): Hyperspectral reflectance properties of plants and potential for classification. *Pest Management Science*, 70(12):1910–1917.

Reusch, S., Jasper, J., and Link, A. (2010). Estimating crop biomass and nitrogen uptake using CropSpec TM , a newly developed activecrop-canopy reflectance sensor. In *In Proceedings of the 10th International Conference on Positron Annihilation (ICPA)*, page 381, Denver, CO, USA.

Robinson, D. E., Soltani, N., Shropshire, C., and Sikkema, P. H. (2013). Cyprosulfamide safens isoxaflutole in sweet corn (Zea mays L.). *HortScience horts*, 48(10):1262 – 1265.

Roth, L. and Streit, B. (2017). Predicting cover crop biomass by lightweight UAS-based RGB and NIR photography: an applied photogrammetric approach. *Precision Agriculture*, pages 1–22.

Rouse, J. W., Haas, R. H., Schell, J. A., and Deering, D. (1973). Monitoring the vernal advancement and retrogradation (green wave effect) of natural vegetation. *NASA, Progress Report RSC 1978-1*, page 112.

Rumpf, T., Mahlein, A.-K., Steiner, U., Oerke, E.-C., Dehne, H.-W., and Plümer, L. (2010). Early detection and classification of plant diseases with support vector machines based on hyperspectral reflectance. *Computers and Electronics in Agriculture*, 74(1):91–99.

Salamí, E., Barrado, C., and Pastor, E. (2014). UAV Flight Experiments Applied to the Remote Sensing of Vegetated Areas. *Remote Sensing*, 6(11):11051–11081.

Sandau, R. (2005). *Digitale Luftbildkamera - Einführung und Grundlagen*. Herbert Wichmann Verlag, Hüthig GmbH & Co. KG, Heidelberg.

Santel, H.-J. (2012). Thiencarbazone-methyl (TCM) and Cyprosulfamide (CSA) - a new herbicide and a new safener for use in corn. In *Nr. 434 (2012): Tagungsband 25. Deutsche Arbeitsbesprechung über Fragen der Unkrautbiologie und -bekämpfung*. Julius Kühn-Institut.

Schepers, J., VanToai, T., Shanahan, J. F., Holland, K. H., Schepers, J. S., Francis, D. D., Schlemmer, M. R., and Caldwell, R. (2003). Use of a Crop Canopy Reflectance Sensor to Assess Corn Leaf Chlorophyll Content. In *Digital Imaging and Spectral Techniques: Applications to Precision Agriculture and Crop Physiology*, pages 135–150. American Society of Agronomy, Crop Science Society of America, and Soil Science Society of America.

Singh, A. K., Ganapathysubramanian, B., Sarkar, S., and Singh, A. (2018). Deep Learning for Plant Stress Phenotyping: Trends and Future Perspectives. *Trends in Plant Science*, 23(10):883 – 898.

Soria-Ruíz, J. and Fernández-Ordoñez, Y. (2003). Prediction of corn yield in Mexico using vegetation indices from NOAA-AVHRR satellite images and degree-days. *Geocarto International*, 18(4):33–42.

Srivastava, N., Hinton, G., Krizhevsky, A., Sutskever, I., and Salakhutdinov, R. (2014). Dropout: a simple way to prevent neural networks from overfitting. *Home Page Papers Submissions News Editorial Board Announcements Proceedings Open Source Software Search Statistics Login Contact Us RSS Feed Journal of Machine Learning Research*, 15(1):1929–1958.

Stafford, J. V. (2000). Implementing Precision Agriculture in the 21st Century. *Journal of Agricultural Engineering Research*, 76(3):267–275.

Streibig, J. C. (1988). Herbicide bioassay. *Weed Research*, 28(6):479–484.

Tavaziva, V. J., Verwijst, T., and Lundkvist, A. (2019). Growth and development of Cirsium arvense in relation to herbicide dose, timing of herbicide application and crop presence. *Acta Agriculturae Scandinavica, Section B — Soil & Plant Science*, 69(3):189–198.

Taylor-Lovell, S., Sims, G. K., Wax, L. M., and Hassett, J. J. (2000). Hydrolysis and soil adsorption of the labile herbicide isoxaflutole. *Environmental Science & Technology*, 34(15):3186–3190.

Testa, G., Reyneri, A., and Blandino, M. (2016). Maize grain yield enhancement through high plant density cultivation with different inter-row and intra-row spacings. *European Journal of Agronomy*, 72:28–37.

Thenkabail, P. S., Lyon, J. G., and Huete, A. (2018a). *Advanced Applications in Remote Sensing of Agricultural Crops and Natural Vegetation*. CRC Press, Boca Raton, USA.

Thenkabail, P. S., Lyon, J. G., and Huete, A. (2018b). *Hyperspectral Indices and Image Classifications for Agriculture and Vegetation*. CRC Press, Boca Raton, USA.

Thorp, K. and Tian, L. (2004). A review on remote sensing of weeds in agriculture. *Precision Agriculture*, 5(5):477–508.

Tian, L., Reid, J. F., and Hummel, J. W. (1999). Development of a precision sprayer for site-specific weed management. 42(4):893–900.

Tilly, N., Aasen, H., and Bareth, G. (2015). Fusion of Plant Height and Vegetation Indices for the Estimation of Barley Biomass. *Remote Sensing*, 7(9):11449–11480.

Tomlin, C. D. S. (2002). The e-pesticide manual, version 2.2. *The e-pesticide manual*, Version 2.2(Ed. 12).

Torra, J., Royo-Esnal, A., Chantre, G. R., and Recasens, J. (2018). Artificial neural networks to model weed emergence. *XVI Congreso de la Sociedad Española de Malherbología, SEMh 2017, Pamplona-Iruña, España, 25-27 octubre, 2017*, pages 63–67.

Torres-Sánchez, J., López-Granados, F., de Castro, A. I., and Peña-Barragán, J. M. (2013). Configuration and Specifications of an Unmanned Aerial Vehicle (UAV) for Early Site Specific Weed Management. *PLoS ONE*, 8(3):e58210.

Torres-Sánchez, J., Peña, J. M., de Castro, A. I., and López-Granados, F. (2014). Multi-temporal mapping of the vegetation fraction in early-season wheat fields using images from UAV. *Computers and Electronics in Agriculture*, 103:104–113.

Turkington, R., Kenkel, N. C., and Franko, G. D. (1980). The biology of canadian weeds: 42 Stellaria media (L.) Vill. *Canadian Journal of Plant Science*, 60(3):981–992.

Turner, D., Lucieer, A., and Watson, C. (2012). An Automated Technique for Generating Georectified Mosaics from Ultra-High Resolution Unmanned Aerial Vehicle (UAV) Imagery, Based on Structure from Motion (SfM) Point Clouds. *Remote Sensing*, 4(12):1392–1410.

Wang, P., Peteinatos, G., Li, H., Brändle, F., Pfündel, E., Drobny, H. G., and Gerhards, R. (2017). Rapid monitoring of herbicide-resistant Alopecurus myosuroides Huds. using chlorophyll fluorescence imaging technology. *Journal of Plant Diseases and Protection*, 125(2):187–195.

Wang, P., Peteinatos, G., Li, H., and Gerhards, R. (2016). Rapid in-season detection of herbicide resistant Alopecurus myosuroides using a mobile fluorescence imaging sensor. *Crop Protection*, 89:170–177.

Weber, J. F., Kunz, C., Peteinatos, G. G., Santel, H.-J., and Gerhards, R. (2017). Utilization of chlorophyll fluorescence imaging technology to detect plant injury by herbicides in sugar beet and soybean. *Weed Technology*, 31(4):523–535.

Weis, M. and Gerhards, R. (2007). Feature extraction for the identification of weed species in digital images for the purpose of site-specific weed control. *Precision Agriculture 2007 - Papers Presented at the 6th European Conference on Precision Agriculture, ECPA 2007*, pages 537–544.

Welton, F. A., Morris, V. H., and Hartzler, A. J. (1929). Organic food reserves in relation to the eradication of Canada thistles. *Ohio, Agricultural Experiment Station*.

Williams II, M. M., Gerhards, R., and Mortensen, D. A. (2000). Two-Year Weed Seedling Population Responses to a Post-Emergent Method of Site-Specific Weed Management. *Precision Agriculture*, 2(3):247–263.

Wilson, B. J., Wright, K. J., Brain, P., Clements, M., and Stephens, E. (1995). Predicting the competitive effects of weed and crop density on weed biomass, weed seed production and crop yield in wheat. *Weed Research*, 35(4):265–278.

Witten, I. H., Frank, E., and Hall, M. A. (2011). *Data mining: practical machine learning tools and techniques*. Morgan Kaufmann, Burlington, USA.

Woebbecke, D. M., Meyer, G. E., Von Bargen, K., and Mortensen, D. A. (1995). Color Indices for Weed Identification Under Various Soil, Residue, and Lighting Conditions. *Transactions of the ASAE*, 38(1):259–269.

Yan Huang, Bruce, L., Koger, T., and Shaw, D. (2001). Analysis of the effects of cover crop residue on hyperspectral reflectance discrimination of soybean and weeds via Haar transform. In *IGARSS. Scanning the Present and Resolving the Future. Proceedings. IEEE 2001 International Geoscience and Remote Sensing Symposium (Cat. No.01CH37217)*, volume 3, pages 1276–1278. IEEE.

Yancheva, S., Georgieva, L., Kostova, M., Halkoglu, P., Dimitrova, M., and Naimov, S. (2016). Plant pigments content as a marker for herbicide abiotic stress in corn (Zea mays L.). *Emirates Journal of Food and Agriculture*, 28(5):332.

Zaman-Allah, M., Vergara, O., Araus, J. L., Tarekegne, A., Magorokosho, C., Zarco-Tejada, P. J., Hornero, A., Albà, A. H., Das, B., Craufurd, P., Olsen, M., Prasanna, B. M., and Cairns, J. (2015). Unmanned aerial platform-based multi-spectral imaging for field phenotyping of maize. *Plant Methods*, 11(1):35.

Zarco-Tejada, P., Catalina, A., González, M., and Martín, P. (2013). Relationships between net photosynthesis and steady-state chlorophyll fluorescence retrieved from airborne hyperspectral imagery. *Remote Sensing of Environment*, 136:247–258.

Zecha, C. W., Peteinatos, G. G., Link, J., and Claupein, W. (2018). Utilisation of ground and airborne optical sensors for nitrogen level identification and yield prediction in wheat. *Agriculture*, 8(6):79.

Zhang, G., Eddy Patuwo, B., and Y. Hu, M. (1998). Forecasting with artificial neural networks: The state of the art. *International Journal of Forecasting*, 14(1):35–62.

Zhang, Q., Xu, F., Lambert, K. N., and Riechers, D. E. (2007). Safeners coordinately induce the expression of multiple proteins and MRP transcripts involved in herbicide metabolism and detoxification in Triticum tauschii seedling tissues. *Proteomics*, 7(8):1261–1278.

Zhou, J., Khot, L. R., Bahlol, H. Y., Boydston, R., and Miklas, P. N. (2016). Evaluation of ground, proximal and aerial remote sensing technologies for crop stress monitoring. *IFAC-PapersOnLine*, 49(16):22–26.